Higher Engineering Science

W. Bolton

Newnes

OXFORD AUCKLAND BOSTON JOHANNESBURG MELBOURNE NEW DELHI

Newnes
An imprint of Butterworth-Heinemann
Linacre House, Jordan Hill, Oxford OX2 8DP
225 Wildwood Avenue, Woburn, MA 01801–2041
A division of Reed Educational and Professional Publishing Ltd

℞ A member of the Reed Elsevier plc group

First published 1999

© W. Bolton 1999

British Library Cataloguing in Publication Data
A catalogue record for this book is available from the British Library

ISBN 0 7506 4235 1

Printed by Martins The Printers, Berwick upon Tweed

FOR EVERY TITLE THAT WE PUBLISH, BUTTERWORTH-HEINEMANN
WILL PAY FOR BTCV TO PLANT AND CARE FOR A TREE.

Contents

Preface

Aims

This book aims to cover the new BTEC mandatory unit Engineering Science for the Higher National Certificates and Diplomas in Engineering. This unit is a broad based unit covering the mechanical principles, electrical principles and engineering system principles underpinning the design and operation of engineering systems and provide the basis for further study in specialist areas of engineering.

The book is likely to contain more material than might appear in any one interpretation of the unit, the aim having been to supply all the material that is likely to appear in any interpretation of the unit.

Structure of the book

The book has been designed to give a clear exposition and guide readers through the scientific principles, reviewing background principles where necessary. Each chapter includes numerous worked examples, self-check revision questions and problems. Answers are supplied to all revision questions and problems.

The book can be considered to consist of four main sections, mirroring those in the BTEC unit:

Static and dynamic systems
  Chapter 1: Structural analysis
  Chapter 2: Bending
  Chapter 3: Torsion
  Chapter 4: Linear and angular motion
  Chapter 5: Mechanical oscillations

Energy transfer
  Chapter 6: Heat transfer
  Chapter 7: Fluid flow

Single phase a.c. theory
  Chapter 8: Single phase a.c. theory
  Chapter 9: Complex numbers
  Chapter 10: Resonant circuits
  Chapter 11: Complex waveforms

Systems

Performance outcomes

The following indicate the outcomes for which each chapter has been planned. At the end of the chapters the reader should be able to:

Chapter 1: Structural analysis
Solve engineering problems involving the axial loading of structures.
Determine the stresses produced in structures as a consequence of changes in temperature.

Chapter 2: Bending
Draw shear force and bending moment diagrams for beams subject to bending.
Determine the stresses arising from bending.

Chapter 3: Torsion
Determine shear stresses and angular deflections arising from the torsion of circular shafts.
Determine the power transmitted by shafts.

Chapter 4: Linear and angular motion
Solve problems involving linear motion with constant acceleration.
Solve problems involving two and three-dimension linear motion.
Solve problems involving angular motion with constant acceleration.
Solve problems involving Newton's laws
Solve problems involving torque and angular motion.
Solve problems involving linear and angular kinetic energy.

Chapter 5: Mechanical oscillations
Explain the principles of simple harmonic motion and solve problems involving such motion.
Describe the effects of damping and the forcing of oscillations.

Chapter 6: Heat transfer
Solve problems involving the transfer of heat by conduction, convection and radiation.

Chapter 7: Fluid flow
Describe the effects of viscosity in fluid flow and solve problems involving fluid flow between parallel plates.
Determine the power losses arising with bearings.
Describe laminar and turbulent flow.
Determine the energy losses arising from the flow of fluids through pipes.

Chapter 8: Single phase a.c. theory

Solve single phase a.c. series and parallel circuit problems by the use of phasors and the use of drawings and simple trigonometry.

Explain and use the terms reactance, susceptance, impedance and admittance.

Determine the power developed in a.c. circuits.

Explain the term power factor, its significance in electrical power transmission, and how it can be improved.

Chapter 9: Complex numbers

Solve problems on series, parallel and series-parallel circuits supplied by a constant sinusoidal voltage by the use of the complex number representation of phasors.

Chapter 10: Resonant circuits

Solve problems on series and parallel resonant circuits.

Chapter 11: Complex waveforms

Describe the nature of complex waveforms in terms of harmonics and synthesise graphically such waveforms.

Describe how electrical and electronic circuit elements can produce such waveforms.

Solve problems involving circuits supplied by a constant complex waveform voltage.

Chapter 12: Systems

Represent systems by block models and analyse such systems.

Describe the methods by which electrical signals convey information.

Describe the characteristics of electronic systems used for analogue-to-digital conversion, digital-to-analogue conversion, amplification and oscillation.

Chapter 13: Control systems

Describe and analyse open-loop and closed-loop control systems in terms of block diagram models.

Describe methods used for electrical and electronic switching of actuators.

Describe methods used for the speed control of motors.

W. Bolton

1 Structural analysis

1.1 Introduction

Load-bearing structures can take many forms. For example, for the building column shown in Figure 1.1(a) the load of the floors and structure above it are applying forces which tend to axially squash the column. For the simple beam bridge in Figure 1.1(b) the load arising from a car crossing it will tend to bend the beam. For the aeroplane in Figure 1.1(c), the lift forces on the wings will tend to bend the wings. For the electric motor in Figure 1.1(d), the load will cause the shaft to become twisted as the motor rotates it, the loading being said to be torsional. To analyse structures so that we can predict their behaviour when loaded it is usual to consider certain basic forms of loading, namely axial tension or compression, bending and torsion.

In this chapter, axial loading is considered, in chapter 2 bending and in chapter 3 torsional loading. Such analysis is necessary for the safe design of structures, whether they be buildings, bridges, aeroplanes, or motors rotating loads.

1.2 Axial loading

Consider a straight bar of constant cross-sectional area when external axial forces are applied at its ends (Figure 1.2(a)). If the forces stretch the bar then the bar is said to be in *tension*, if they compress it in *compression*.

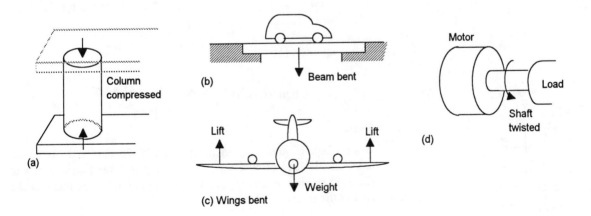

Figure 1.1 *Examples of loading*

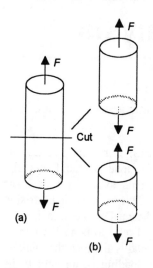

Figure 1.2 *(a) Bar in tension, and (b) with an imaginary sectional cut*

1.2.1 Direct stress

It is necessary in analysing structures to be able to distinguish between external forces, such as applied loads, and internal forces which are produced in structural members as a result of applying the loads. Consider an axially loaded bar. We can think of the bar as being like a spring which, when stretched by external forces F, sets up internal forces which resist the external forces extending it. Consider a plane in the bar which is at right angles to its axis and suppose we make an imaginary cut along that plane (Figure 1.2(b)). Equal forces F are required at the break to maintain equilibrium of the two lengths of the bar. This is true for any section across the bar and hence there is a force F acting on any imaginary section perpendicular to the axis of the bar. Thus we can consider, in this case, that there are internal forces F across any section at right angles to the axis. Internal forces are responsible for what is termed *stress*.

The term *direct stress* is used for the value of this internal force per unit area of the plane, with the stresses being termed *tensile stresses* if the forces are in such directions as to stretch the bar and *compressive stresses* if they compress the bar. It is customary to denote tensile stresses as positive and compressive stresses as negative. The symbol used for the stress is σ and, it is defined by:

$$\sigma = \frac{F}{A} \qquad\qquad\qquad [1]$$

where A is the cross-sectional area. It has the units of force per unit area. The unit of N/m² is termed the pascal (Pa).

Example

A bar with a uniform rectangular cross-section of 20 mm by 40 mm is subjected to an axial force of 50 kN. Determine the tensile stress in the bar.

Using equation [1]:

$$\sigma = \frac{F}{A} = \frac{50 \times 10^3}{0.020 \times 0.040} = 62.5 \times 10^6 \text{ Pa} = 62.5 \text{ MP}$$

Example

A steel bolt (Figure 1.3) has a diameter of 25 mm and carries an axial tensile load of 50 kN. Determine the average tensile stress at the shaft section aa and the screwed section bb if the diameter at the root of the thread is 21 mm.

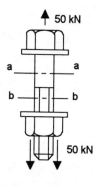

Figure 1.3 *Example*

The average tensile stress at section aa is given by equation [1] as:

$$\sigma_{aa} = \frac{50 \times 10^3}{\frac{1}{4}\pi 0.025^2} = 101.9 \times 10^6 \text{ Pa} = 101.9 \text{ MP}$$

The average tensile stress at section bb is given by equation [1] as:

$$\sigma_{bb} = \frac{50 \times 10^3}{\frac{1}{4}\pi 0.021^2} = 144.4 \times 10^6 \text{ Pa} = 144.4 \text{ MP}$$

Revision

1   A circular cross-section rod with a diameter of 20 mm is stretched by axial forces of 10 kN. What is the stress in the bar?

2   A circular cross-section column with a diameter of 100 mm is compressed by axial forces of 70 kN. What is the stress in the column?

1.2.2 Direct strain

An axial loaded bar undergoes a change in length, increasing in length when in tension (Figure 1.4) and decreasing in length when in compression. The change in length e for an original length L is termed the *direct strain* ε.

$$\varepsilon = \frac{e}{L} \qquad\qquad [2]$$

Since strain is a ratio of two lengths it is a dimensionless number, i.e. it has no units.

When the change in length is an increase in length then the strain is termed *tensile strain* and is positive. When the change in length is a decrease in length then the strain is termed *compressive strain* and is negative.

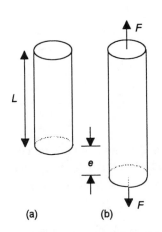

Figure 1.4   *(a) Unstretched bar, (b) stretched bar*

Example

Determine the strain experienced by a rod of length 100.0 cm when it is stretched by 0.2 cm.

Using equation [2]:

$$\text{strain} = \frac{e}{L} = \frac{0.2}{100} = 0.0002$$

Revision

3   A column of height 40 cm contracts axially by 0.02 cm when loaded. What is the strain experienced by the column?

1.2.3 Hooke's law

Hooke's law states that strain is proportional to the stress producing it (Figure 1.5). This law can generally be assumed to be obeyed within certain limits of stress by most of the metals used in engineering. Within

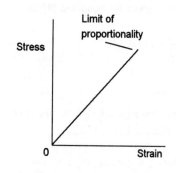

Figure 1.5   *Hooke's law*

the limits to which Hooke's law is obeyed, the ratio of the direct stress to the strain produced is called the *modulus of elasticity E*:

$$E = \frac{\sigma}{\varepsilon} \qquad \qquad [3]$$

For a bar of uniform cross-sectional area A and length L, subject to axial force F and extending by e, equation [3] becomes:

$$E = \frac{FL}{Ae} \qquad \qquad [4]$$

Example

A circular cross-section steel bar of uniform diameter 12 mm and length 1 m is subject to tensile forces of 10 kN. If the steel has a modulus of elasticity of 200 GPa, what will be the stress and strain in the bar?

Using equation [1]:

$$\text{stress} = \frac{F}{A} = \frac{10 \times 10^3}{\frac{1}{4}\pi 0.012^2} = 88.4 \times 10^6 \text{ Pa} = 88.4 \text{ MP}$$

Using equation [3]:

$$\text{strain} = \frac{\sigma}{E} = \frac{88.4 \times 10^6}{200 \times 10^9} = 4.42 \times 10^{-4}$$

Revision

4   A rod has a uniform cross-sectional area of 400 mm² and a length of 1.6 m. Determine the elongation of the rod when it is subject to axial tensile forces of 28 kN if the material has a modulus of elasticity of 200 GPa.

5   A mass of 6 kg is suspended by a vertical steel wire, modulus of elasticity 200 GPa, from a beam. If the wire has a uniform diameter of 2.5 mm and a length of 5 m, by how much will the wire stretch? Ignore the weight of the wire.

1.3 Axially loaded members

Consider two forms of axially loaded members, one in which a bar is formed by combining two members in parallel and one in which they are combined in series.

1.3.1 Members in parallel

Tensile or compressive members which consist of two or more bars or tubes in parallel are termed *compound bars*. Figure 1.6 shows such an arrangement involving a central rod A of one material in a tube B of

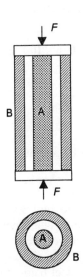

Figure 1.6 *Example of a compound bar*

another material, the load being applied to plates fixed across the tube ends so that the load is applied to both A and B.

With such a compound bar, the load F applied is shared by the members. Thus if F_A is the force acting on member A and F_B is the force acting on member B, for *equilibrium* we must have:

$$F_A + F_B = F \qquad [5]$$

If σ_A is the resulting stress in element A and A_A is its cross-sectional area, $\sigma_A = F_A/A_A$. Likewise, if σ_B is the stress in element B and A_B is its cross-sectional area, $\sigma_B = F_B/A_B$. Thus the equilibrium equation [5] can be written as:

$$\sigma_A A_A + \sigma_B A_B = F \qquad [6]$$

Since the elements A and B are the same initial length and must remain together when loaded, the strain in A of ε_A must be the same as that in B of ε_B. Thus, for *compatibility*, we have:

$$\varepsilon_A = \varepsilon_B \qquad [7]$$

Thus, using Hooke's law, we must have:

$$\frac{\sigma_A}{E_A} = \frac{\sigma_B}{E_B} \qquad [8]$$

where E_A is the modulus of elasticity of the material of element A and E_B that of the material of element B.

Example

A compound bar consists of a brass rod of diameter 30 mm inside a cylindrical steel tube of internal diameter 35 mm. What should be the external diameter of the steel tube if the stress in the brass rod is not to exceed 80 MPa when the compound bar is subject to an axial load of 200 kN. The modulus of elasticity for the steel is 200 GPa and that for the brass 120 GPa.

Using equation [8]:

$$\frac{\sigma_S}{E_S} = \frac{\sigma_B}{E_B}$$

$$\sigma_S = \frac{80 \times 10^6 \times 200 \times 10^9}{120 \times 10^9} = 133.3 \times 10^6 \, P$$

Using equation [6]:

$$F = \sigma_S A_S + \sigma_B A_B = \sigma_S A_S + 80 \times 10^6 \times \tfrac{1}{4}\pi 0.030^2 = 200 \times 10^3 \, N$$

Hence substituting the value of stress obtained above:

$$A_S = \frac{200 \times 10^3 - 80 \times 10^6 \times \frac{1}{4}\pi 0.030^2}{133.3 \times 10^6} = 1.076 \times 10^{-3} \text{ m}$$

Thus:

$$1.076 \times 10^{-3} = \tfrac{1}{4}\pi D^2 - \tfrac{1}{4}\pi 0.035^2$$

and D, the external diameter of the steel tube is 0.0509 m.

Revision

6   A reinforced concrete column is uniformly 500 mm square and consists of four steel rods, each of diameter 25 mm, embedded in the concrete (Figure 1.7). Determine the compressive stresses in the concrete and the steel when the column is subject to a compressive load of 1 MN, the modulus of elasticity of the steel being 200 GPa and that of the concrete 14 GPa.

7   A compound bar of length 500 mm consists of a steel rod of diameter 20 mm in a brass tube of internal diameter 20 mm and external diameter 30 mm. Determine the stress in each material and the extension of the bar when axial tensile forces of 30 kN are applied. The modulus of elasticity for the steel is 200 GPa and that for the brass 90 GPa.

8   A compound beam consists of a square cross-section timber core 75 mm by 75 mm with steel plates 75 mm by 12 mm bolted along its entire length to opposite faces (Figure 1.8). Determine the maximum permissible axial tensile load if the maximum permissible stress in the timber is 6.3 MPa and in the steel 140 MPa, the modulus of elasticity of the timber being 8 GPa and for the steel 200 GPa.

9   A copper rod of diameter 25 mm is inserted into a steel tube of internal diameter 35 mm and external diameter 40 mm, the rod and tube being attached at each end. Determine the stresses in the rod and tube when the compound arrangement is subject to an axial tensile load of 40 kN, the modulus of elasticity of the steel being 200 GPa and that of the copper 95 GPa.

1.3.2 Members in series

Consider a composite bar consisting of two, or more, members in series. These may be of different materials and/or different cross-sections. Figure 1.9 illustrates the type of situation that might occur. Here we have three rods connected end-to-end, the rods perhaps being of different cross-sections and perhaps materials. The composite is subject to a tensile load.

With just a single load we must have, for equilibrium, the same forces acting on each of the series members. Thus the forces stretching member

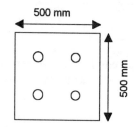

Figure 1.7   *Revision problem 6*

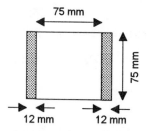

Figure 1.8   *Revision problem 8*

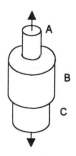

Figure 1.9   *Members in series*

A are the same as those stretching member B and the same as those stretching member C. The extensions of the members may, however, differ. The total extension of the composite bar will be the sum of the extensions arising for each series element.

Example

Steel rods of diameter 10 mm and 15 mm are connected to either end of a copper bar of diameter 20 mm (a similar situation to that shown in Figure 1.9). The 10 mm diameter steel rod has a length of 600 mm, the 15 mm rod a length of 400 mm and the copper rod a length of 800 mm. Determine the stresses in each of the rods and the total elongation if the composite is subject to axial tensile forces of 12 kN. The steel has a modulus of elasticity of 200 GPa and the copper one of 100 GPa.

The forces acting axially on the 10 mm steel rod are 12 kN and thus the stress in that member is:

$$\sigma = \frac{12 \times 10^3}{\frac{1}{4}\pi 0.010^2} = 152.8 \text{ MPa}$$

Assuming Hooke's law, this rod will extend by:

$$e = L \times \frac{\sigma}{E} = \frac{0.600 \times 152.8 \times 10^6}{200 \times 10^9} = 0.000\,458$$

Likewise the forces acting on the copper rod are 12 kN and so the stress in it is:

$$\sigma = \frac{12 \times 10^3}{\frac{1}{4}\pi 0.020^2} = 38.2 \text{ MPa}$$

and it extends by:

$$e = \frac{0.800 \times 38.2 \times 10^6}{100 \times 10^9} = 0.000\,306$$

The forces acting on the other steel rod are also 12 kN. Thus the stress in it is:

$$\sigma = \frac{12 \times 10^3}{\frac{1}{4}\pi 0.015^2} = 67.9 \text{ MPa}$$

and its extension is:

$$e = \frac{0.400 \times 67.9 \times 10^6}{200 \times 10^9} = 0.000\,136$$

The total extension is the sum of the three extensions and so is 0.458 + 0.306 + 0.136 = 0.900 mm.

Revision

10 A steel bar with a total length of 240 mm has a diameter of 40 mm for a length of 100 mm, a diameter of 30 mm for a length of 60 mm and a diameter of 20 mm for the remaining 80 mm of its length. Determine the tensile load required to produce a total elongation of 0.177 mm for the rod. The modulus of elasticity is 200 GPa.

11 A steel rod with a diameter of 12 mm and a length of 3 m is joined to the end of an aluminium rod of diameter 12 mm and length 2 m. Determine the overall extension of the rod when it is subject to an axial tensile load of 18 kN. The modulus of elasticity of the steel is 200 GPa and that of the aluminium 70 GPa.

1.4 Poisson's ratio

When a material is longitudinally stretched it contracts in a transverse direction (Figure 1.10). The ratio of the transverse strain to the longitudinal strain is called *Poisson's ratio*.

$$\text{Poisson's ratio} = -\frac{\text{transverse strain}}{\text{longitudinal strain}} \qquad [9]$$

The minus sign is because when one of the strains is tensile the other is compressive. For most engineering metals, Poisson's ratio is about 0.3.

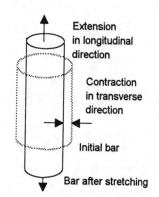

Figure 1.10 *Transverse contraction as a result of longitudinal stretching*

Example

A steel bar of length 1 m is extended by 0.1 mm. By how much will the width of the bar contract if initially the width was 100 mm? Poisson's ratio is 0.31.

The longitudinal strain is 0.1/1000 = 0.000 1. Thus, using equation [9], the transverse strain is:

transverse strain = $-0.31 \times 0.000\ 1 = -3.1 \times 10^{-5}$

Thus:

change in width = original width × transverse strain

$$= 100 \times (-3.1 \times 10^{-5}) = -3.1 \times 10^{-3} \text{ mm}$$

The minus sign indicates that the width is reduced by this amount.

Revision

12 A steel bar with a rectangular cross-section 75 mm by 25 mm is subject to a tensile longitudinal load of 200 kN. Determine the decrease in the lengths of the sides of the resulting cross-section.

The material has an elastic modulus of 200 GPa and Poisson's ratio of 0.3.

1.5 Temperature stresses

When the temperature of a body changes it changes in length. If this expansion or contraction is wholly or partially resisted, stresses are set up in the body. Consider a bar of material of initial length L_0. If the temperature is now raised to θ and the bar is free to expand, the length increases to $L_0(1 + a\theta)$, where a is the coefficient of linear expansion of the bar material. The change in length of the bar is thus $L_0(1 + a\theta) - L_0 = L_0 a\theta$. If this expansion is prevented, it is as if a bar of length $L_0(1 + a\theta)$ has been compressed to a length L_0. The compressive strain ε is thus:

$$\varepsilon = \frac{L_0 a\theta}{L_0(1 + a\theta)}$$

Since $a\theta$ is small compared with 1, the strain ε can be written as:

$$\varepsilon = a\theta \qquad\qquad [10]$$

If the material has a modulus of elasticity E then, assuming Hooke's law is obeyed, the stress σ produced is:

$$\sigma = a\theta E \qquad\qquad [11]$$

The stress is independent of the length and cross-section of the bar.

Example

Determine the stress produced per degree change in temperature for a fully restrained steel member if the coefficient of linear expansion for steel is 12×10^{-6} per °C and the modulus of elasticity is 200 GPa.

Using equation [11]:

$$\sigma = a\theta E = 12 \times 10^{-6} \times 1 \times 200 \times 10^9 = 2.4 \times 10^6 \text{ Pa} = 2.4 \text{ MPa}$$

Example

Determine the minimum gap to be left between crane running lines of steel, length 10 m, if the increase in stress in the steel is not to exceed 24 MPa for a temperature increase of 30°C from when they were set down. The coefficient of linear expansion for steel is 12×10^{-6} per °C and the modulus of elasticity is 200 GPa.

We need to determine the temperature change which will result in a stress of 24 MPa for fully constrained rails. Using equation [11]:

$$\sigma = a\theta E = 12 \times 10^{-6} \times \theta \times 200 \times 10^9 = 24 \times 10^6$$

Hence a temperature change of 10°C will produce such a stress. Thus there must be sufficient gap between the rails for them to expand unconstrained for 20°C. The length L after a 20°C change will be:

$$L = L_0(1 + a\theta) = 10(1 + 12 \times 10^{-6} \times 20) = 10.002\ 4\ \text{m}$$

Hence the gap that should be left is 2.4 mm.

Example

A steel wire is stretched between two rigid supports so that that the wire is subject to a stress of 30 MPa at 20°C. Determine the stress in the wire when the temperature drops to 0°C. The coefficient of linear expansion for steel is 12×10^{-6} per °C and the modulus of elasticity is 200 GPa.

Consider initially just the effect of the drop in temperature. Using equation [11]:

$$\sigma = a\theta E = 12 \times 10^{-6} \times 20 \times 200 \times 10^9 = 48\ \text{MPa}$$

This will be a tensile stress. The total stress acting on the wire will be the sum of the thermal stress and the initial stress, i.e. 78 MPa.

Revision

13 Determine the stress produced per degree change in temperature for a fully restrained aluminium member if the coefficient of linear expansion for aluminium is 22×10^{-6} per °C and the modulus of elasticity is 74 GPa.

14 A brass bar is heated to 60°C and then clamped at its ends. What will be the stress in the bar when it is cooled down to room temperature of 20°C. The coefficient of linear expansion for brass is 18×10^{-6} per °C and the modulus of elasticity 90 GPa.

15 A steel bar is clamped rigidly at both ends at 10°C and subject to a tensile stress of 20 MPa. What will be the stress in the bar if the temperature is now raised to 20°C? The coefficient of linear expansion of the steel is 12×10^{-6} per °C and the modulus of elasticity 200 GPa. Hint: the expansion of the rod will subject it to compressive stresses.

16 A steel beam has a length of 25 m between its supports. It is rigidly fixed at one support but the other support can accommodate 5 mm of expansion movement. Determine the stress produced in the beam when the temperature rises by 40°C. The coefficient of linear expansion for steel is 12×10^{-6} per °C and the modulus of elasticity is 200 GPa.

1.5.1 Compound bars

Consider a compound bar (Figure 1.11(a)) consisting of two members A and B, say a circular bar inside a circular tube, with the two materials having different coefficients of expansion, a_A and a_B, and different modulus of elasticity values, E_A and E_B. The two members are of the same initial length L and attached rigidly together. What are the stresses produced in the materials when the temperature changes by θ? Suppose, however, the two members had not been fixed to each other. When the temperature changes they would have expanded to give the situation illustrated by Figure 1.11(b). A would have changed its length by $La_A\theta_A$ and B its length by $La_B\theta_B$. There will now be a difference in length between the two members at temperature θ of $(a_A - a_B)\theta L$. Thus when the two members are rigidly fixed together (Figure 1.11(c)), this difference in length is eliminated by compressing member B with a force F and extending A with a force F. The extension e_A of A due to this force is:

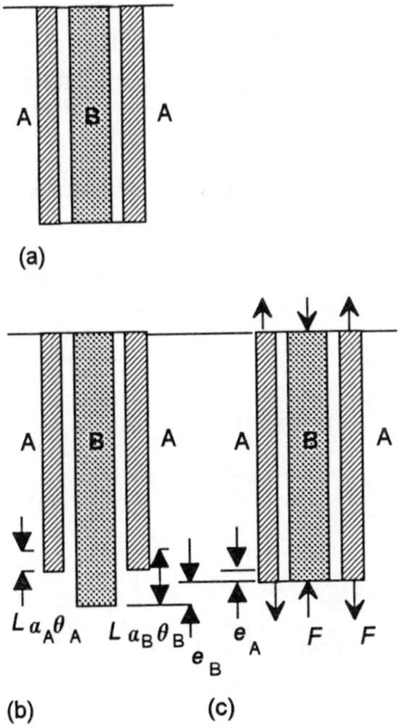

$$e_A = \frac{FL}{E_A A_A}$$

where A_A is its cross-sectional area. The contraction e_B of B due to this force is:

$$e_B = \frac{FL}{E_B A_B}$$

where A_B is its cross-sectional area. But $e_A + e_B = (a_A - a_B)\theta L$. Thus:

$$(a_A - a_B)\theta L = FL\left(\frac{1}{E_A A_A} + \frac{1}{E_B A_B}\right)$$

and so:

$$F = \frac{(a_A - a_B)\theta}{\left(\dfrac{1}{E_A A_A} + \dfrac{1}{E_B A_B}\right)} \qquad [12]$$

If such a compound bar is also subject to loading we can use the *principle of superposition*. This states that the resultant stress or displacement at a point in a bar subject to a number of loads can be determined by finding the stress or displacement caused by each load considered acting separately on the bar and then adding the contributions caused by each load to obtain the resultant stress. Thus if a compound bar is acted on by, say, tensile loads and also its temperature is raised, the stress in a member is the sum of the stresses obtained by considering the thermal stress and the loading separately. This is illustrated in a following example.

(a)

(b)   (c)

Figure 1.11   *Compound bar*

Example

A steel rod of length 1.0 m and 30 mm diameter fits centrally inside a 1.0 m length of copper tubing, it having an internal diameter of 35 mm and an external diameter of 60 mm. The rod and tube are rigidly fixed together at each end. What will be the stresses produced in each by a temperature increase of 100°C? The copper has a modulus of elasticity of 100 GPa and a coefficient of linear expansion of 20×10^{-6} per °C, the steel a modulus of elasticity of 200 GPa and a coefficient of linear expansion of 12×10^{-6} per °C.

Using equation [12], the force compressing the copper and the force extending the steel is:

$$F = \frac{(a_A - a_B)\theta}{\left(\dfrac{1}{E_A A_A} + \dfrac{1}{E_B A_B}\right)}$$

$$= \frac{(20 - 12) \times 10^{-6} \times 100}{\dfrac{1}{100 \times 10^9 \times \frac{1}{4}\pi(0.060^2 - 0.035^2)} + \dfrac{1}{200 \times 10^9 \times \frac{1}{4}\pi 0.030}}$$

$$= 64.3 \text{ kN}$$

The compressive stress acting on the copper is thus:

$$\sigma_A = \frac{64.3 \times 10^3}{\frac{1}{4}\pi(0.060^2 - 0.035^2)} = 34.5 \text{ MPa}$$

The tensile stress acting on the steel is:

$$\sigma_B = \frac{64.3 \times 10^3}{\frac{1}{4}\pi 0.030^2} = 91.0 \text{ MPa}$$

Example

If the heated compound bar in the above example is then subject to a compressive axial load of 50 kN, what will be the stresses in the copper and steel elements.

Considering just the effects of the 50 kN force, equation [8] gives:

$$\frac{\sigma_A}{100 \times 10^9} = \frac{\sigma_B}{200 \times 10^9}$$

Hence $2\sigma_A = \sigma_B$. Equation [6] gives $\sigma_A A_A + \sigma_B A_B = F$ and so:

$$\sigma_A \times \tfrac{1}{4}\pi(0.060^2 - 0.035^2) + \sigma_B \times \tfrac{1}{4}\pi 0.030^2 = 50 \times 10^3$$

$$\sigma_A \times \tfrac{1}{4}\pi(0.060^2 - 0.035^2) + 2\sigma_A \times \tfrac{1}{4}\pi 0.030^2 = 50 \times 10^3$$

The compressive stress in the copper due to the load is 15.2 MPa and the compressive stress in the steel due to the load is 30.4 MPa. Thus the resultant stress, taking into account the thermal stresses, is for the copper a compressive stress of $-15.2 - 34.5 = -49.7$ MPa and for the steel a tensile stress of $-30.4 + 91.0 = 60.6$ MPa.

Revision

17 A compound tube has a length of 750 mm and is fixed between two rigid supports. It consists of a copper tube of external diameter 100 mm and internal diameter 87 mm encasing a steel tube of external diameter 87 mm and internal diameter 75 mm. Determine the stresses set up in the tubes as a result of the temperature being increased by 40°C. The steel has a modulus of elasticity of 210 GPa and a coefficient of linear expansion of 12×10^{-6} per °C and the copper a modulus of elasticity of 130 GPa and a coefficient of linear expansion of 17×10^{-6} per °C.

18 A brass rod of diameter 25 mm is enclosed centrally in a steel tube with internal diameter 25 mm and external diameter 40 mm, both having a length of 1.0 m and rigidly fastened at the ends. Determine the stresses in the rod and tube resulting from a temperature increase of 100°C. The steel has a modulus of elasticity of 200 GPa and a coefficient of linear expansion of 12×10^{-6} per °C and the brass a modulus of elasticity of 100 GPa and a coefficient of linear expansion of 19×10^{-6} per °C.

19 A compound rectangular cross-section bar of length 1.5 m consists of a rectangular cross-section steel bar 50 mm by 10 mm sandwiched between two rectangular cross-section copper bars, each being 50 mm by 10 mm. The three bars are fixed together at the ends. Determine the stresses resulting in the materials due to the temperature being raised by 60°C and the bar being subject to a tensile load of 40 kN. The steel has a modulus of elasticity of 200 GPa and a coefficient of linear expansion of 12×10^{-6} per °C and the copper a modulus of elasticity of 110 GPa and a coefficient of linear expansion of 17×10^{-6} per °C.

1.5.2 Thermal shock

When the temperature changes, an unconstrained material will expand or contract. However, if it is constrained so that its natural expansion or contraction is prevented, then stresses are set up in the material. Suppose we pour hot water into a cold glass. The surface layer immediately in contact with the hot water will rapidly come up to the temperature of the hot water. However, because glass has a low thermal conductivity, the layers of glass under the surface will still be at their original cold temperature. The expansion of the hot surface layer is thus constrained by the underlying layers and so stresses are set up. The shock of putting the hot water in the cold glass can produce stresses which are high enough to fracture the glass.

Problems

1 A circular cross-section rod with a diameter of 25 mm and initial length 500 mm is subject to axial forces of 50 kN which cause it to extend by 0.25 mm. Determine the stress and strain in the rod.

2 A circular cross-section rod with a diameter of 25 mm and initial length 15 m is subject to axial forces of 80 kN. The material has a modulus of elasticity of 200 GPa. Determine the stress and strain in the rod.

3 A hollow circular column of length 1.5 m has an outside diameter of 300 mm and a wall thickness of 25 mm. The column material has a modulus of elasticity of 200 GPa. Determine the compressive stress in the column and its change in length when it carries a compressive load of 700 kN.

4 An aluminium wire of length 3 m is subject to a tensile stress of 70 MPa. If the aluminium has a modulus of elasticity of 70 GPa, determine the elongation of the wire.

5 A bar has a uniform cross-sectional area of 50 mm² and a length of 5 m. What will be its elongation when subject to axial tensile forces of 40 kN if the bar material has a modulus of elasticity of 200 GPa?

6 A hollow circular cross-section column has a length of 600 mm, an external diameter of 75 mm and a wall thickness of 7.5 mm. Determine the change in length of the cylinder when it is subject to axial compressive forces of 50 kN, the material having a modulus of elasticity of 100 GPa.

7 An electrical conductor consists of a 5 mm diameter steel wire coated with copper to give an external diameter of 7 mm. Determine the stresses in the two materials when a length of the conductor is subject to tensile forces of 2 kN. The modulus of elasticity of the steel is 200 GPa and that of the copper 120 GPa.

8 A reinforced concrete column has a 450 mm square uniform cross-section and contains four steel bars, each of diameter 25 mm. Determine the stresses in the steel and the concrete when the column is subject to an axial compressive load of 1.5 MN, the modulus of elasticity of the steel being 200 GPa and that of the concrete 14 GPa.

9 A compound beam is made by sandwiching a steel plate 150 mm by 6 mm by timber members 150 mm by 75 mm (Figure 1.12), the three members being bolted together. The maximum permissible stress for the timber is 6 MPa and for the steel 130 MPa, the modulus of elasticity for the timber being 8.2 GPa and for the steel 205 GPa. Determine the maximum permissible tensile load for the compound beam.

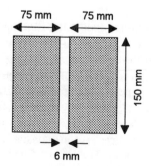

Figure 1.12   *Problem 9*

10 A component consists of a steel bar of diameter 40 mm and length 200 mm with a copper rod of length 400 mm joined to its end. Determine the diameter necessary for the copper rod if the extension of each of the constituent rods is to be the same when the component is subject to an axial load. The steel has a modulus of elasticity of 200 GPa and the copper 100 GPa.

11 A bar of length 3 m and with a circular cross-section of diameter 30 mm is stretched by tensile forces of 85 kN. Determine the elongation of the bar and the decrease in diameter. The material has an elastic modulus of 200 GPa and Poisson's ratio of 0.3.

12 Determine the increase in stress that will occur for a restrained steel beam when subject to a rise in temperature from 10°C to 38°C. The coefficient of linear expansion of the steel is 12×10^{-6} per °C and the modulus of elasticity 200 GPa.

13 A steel bar of length 200 mm is placed between two rigid supports so that there is a clearance of 0.3 mm. Determine the stress produced in the bar if its temperature is raised by 200°C. The coefficient of linear expansion of the steel is 12×10^{-6} per °C and the modulus of elasticity 200 GPa.

14 A railroad has steel rails of length 12 m with a clearance of 3 mm at a temperature of 10°C. Determine the temperature at which the rails will touch each other and determine what the stress would have been at this temperature if there had been no clearance. The coefficient of linear expansion of the steel is 12×10^{-6} per °C and the modulus of elasticity 200 GPa.

15 A copper rod of diameter 15 mm is enclosed centrally in a steel tube with internal diameter 18 mm and external diameter 24 mm, both having a length of 1.0 m and rigidly fastened at the ends. Determine the stresses in the rod and tube resulting from a temperature increase of 100°C. The steel has a modulus of elasticity of 200 GPa and a coefficient of linear expansion of 12×10^{-6} per °C and the copper a modulus of elasticity of 100 GPa and a coefficient of linear expansion of 18×10^{-6} per °C.

16 A solid aluminium cylinder with a cross-sectional area of 6000 mm² is centrally located inside a steel tube with material of cross-sectional area 2000 mm², both being the same length of 500 mm. A compressive load of 200 kN is axially applied and the temperature increased by 60°C. Determine the stresses carried by the cylinder and the tube. The steel has a modulus of elasticity of 210 GPa and a coefficient of linear expansion of 12×10^{-6} per °C and the aluminium a modulus of elasticity of 70 GPa and a coefficient of linear expansion of 23×10^{-6} per °C.

2 Bending

2.1 Introduction

As discussed in chapter 1 in relation to possible forms of loading structures, one basic form involves bending. Thus, for the simple beam bridge in Figure 2.1(a) the load arising from a car crossing it will tend to bend the beam and for the aeroplane in Figure 2.1(b) the lift forces on the wings will tend to bend the wings. This chapter is a discussion of loading due to bending, the various forms it can take, and the stresses that can arise from such bending.

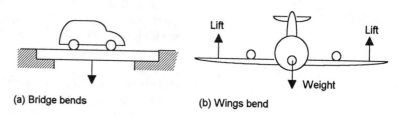

(a) Bridge bends   (b) Wings bend

Figure 2.1 *Examples of structures where bending occurs*

2.1.1 Beams

A *beam* can be defined as a structural member to which loads are applied and which cause it to bend. When beams bend they become curved. The following are some examples of types of beams:

1   *Cantilever* (Figure 2.2(a))
    This is a beam which is rigidly fixed at just one end. This rigid fixing prevents rotation of the beam when a load is applied to the cantilever. Thus if you apply a vertical force some distance from the fixed end there will be a moment which, for equilibrium, has to be balanced by a resisting moment at the fixed end.

2   *Simply supported beam* (Figure 2.2(b))
    This is a beam which is supported at its ends on rollers or smooth surfaces or one of these combined with a pin at the other end.

3   *Simply supported beam with overhanging ends* (Figure 12.2(c))
    This is a simple supported beam with the supports set in some distance from the ends.

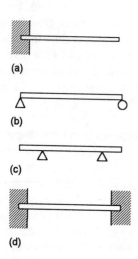

(a)

(b)

(c)

(d)

Figure 2.2 *Examples of beams*

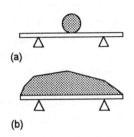

(a)

(b)

Figure 2.3   Loads: (a) concentrated, (b) distributed

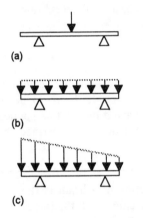

(a)

(b)

(c)

Figure 2.4   Loads: (a) concentrated, (b) uniformly distributed, (c) non-uniformly distributed

4   *Built-in beam* (Figure 2.2(d))
This is a beam which is built-in at both ends and so both ends are rigidly fixed.

Where an end is rigidly fixed there is a reaction force and a resisting moment. At a supported end or point there are reactions but no resisting moments. At a free end there are no reactions and no resisting moments.

2.1.2 Loads

The loads that can be carried by beams may be concentrated or distributed. A concentrated load is one which can be considered to be applied at a point (Figure 2.3(a)); a distributed load is one that is applied over a significant length of the beam (Figure 2.3(b)). An obvious example of a distributed load is the weight of the beam, there being a weight force for each unit length of the beam. On figures, concentrated loads are represented by single arrows acting along the line concerned (Figure 2.4(a)); distributed loads are represented by a series of arrows along the length of beam over which the load is distributed (Figure 2.4(b)). With a uniformly distributed load the arrows are all the same length; if the distributed load is not uniform then the lengths of the arrows are varied to indicate how the distributed load is varying (Figure 2.4(c)). The loading on a beam may be a combination of fixed loads and distributed loads.

2.1.3 Types of beams

Beams can take a number of forms (Figure 2.5). For example, they might be simple rectangular sections, e.g. the timber joists used in the floor construction of houses, circular sections, tubes, e.g. tubes carrying liquids and supported at a number of points, and the very widely used *universal beam*. The universal beam is an I-section and such beams are widely available from stockists in a range of sizes and weights.

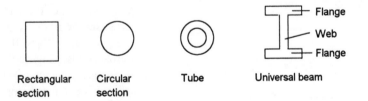

Rectangular section   Circular section   Tube   Universal beam

Figure 2.5   *Forms of beams*

2.2 Shear force and bending moment

The stresses in a beam when bent are more complex than those occurring in a bar which is subject to longitudinal forces and is in simple tension or compression. In order to evaluate the stresses within a bent beam at a particular point, we need to consider the shear force and bending moment at the point.

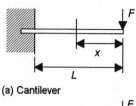

(a) Cantilever

(b) Free body diagram

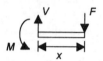

(c) Forces for vertical equilibrium

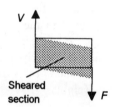

(d) Vertical and moment equilibrium

Figure 2.6   *Cantilever*

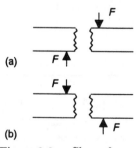

Figure 2.7   *Shear*

(a)

(b)

Figure 2.8   *Shear force:*
(a) positive, (b) negative

Consider a cantilever (Figure 2.6(a) which has a concentrated load F applied at the free end and an imaginary cut through the beam at a distance x from the free end . We will think of the cut section of beam as a *free body*, isolated from the rest of the beam and effectively floating in space. Now consider the conditions for the equilibrium of the section of beam to the right of the cut (Figure 2.6(b)).

For the section of beam to be in vertical equilibrium, we must have a vertical force V acting on it such that $V = F$ (Figure 2.6(c)). This force V is called the *shear force* because the combined action of V and F on the section is to shear it (Figure 2.7). In general:

> The shear force at a transverse section of a beam is the algebraic sum of the external forces acting at right angles to the axis of the beam on one side of the section concerned.

In addition to vertical equilibrium we must also have the section of beam in rotational equilibrium. For the section of the beam to be in moment equilibrium and not rotate, we must have a moment M applied (Figure 2.6(d)) at the cut so that $M = Fx$. This moment is termed the *bending moment*.

> The bending moment at a transverse section of a beam is the algebraic sum of the moments about the section of all the forces acting on one (either) side of the section concerned.

The conventions most often used for the signs of shear forces and bending moments are:

1   *Shear force*
When the shear forces on either side of a section are clockwise (Figure 2.8(a)), i.e. the left-hand side of the beam is being pushed upwards and the right-hand side downwards, the shear force is taken as being positive. When the shear forces on either side of a section are anticlockwise (Figure 2.8(b)), i.e. the left-hand side of the beam is being pushed downwards and the right-hand side upwards, the shear force is taken as being negative.

2   *Bending moment*
Bending moments are positive if they give rise to sagging (Figure 2.9(a) and negative if they give rise to hogging (Figure 2.9(b)).

Example

Determine the shear force and bending moment at points 1 m and 4 m from the right-hand end of the beam shown in Figure 2.10. Neglect the weight of the beam.

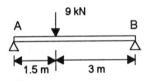

(a) Sagging

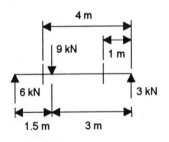

(b) Hogging

Figure 2.9   *Bending moment:*
(a) positive, (b) negative

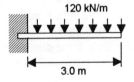

Figure 2.10   *Example*

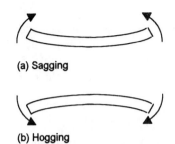

Figure 2.11   *Example*

The reactions at the ends A and B can be found by taking moments about A:

$$R_B \times 4.5 = 9 \times 1.5$$

to give $R_B = 3$ kN and then considering the vertical equilibrium which gives:

$$R_A + R_B = 9$$

and thus $R_A = 6$ kN. Figure 2.11 shows the forces acting on the beam.

If we make an imaginary cut in the beam at 1 m from the right-hand end, then the force on the beam to the right of the cut is 3 kN upwards and that to the left is $9 - 6 = 3$ kN downwards. The shear force is thus negative and -3 kN.

If we make an imaginary cut in the beam at 4 m from the right-hand end, then the force on the beam to the right of the cut is $9 - 3 = 6$ kN downwards and that to the left is 6 kN upwards. The shear force is thus positive and $+6$ kN.

The bending moment at a distance of 1 m from the right-hand end of the beam, when we consider that part of the beam to the right, is 3×1 kN m. Since the beam is sagging the bending moment is $+3$ kN m. At a distance of 4 m from the right-hand end of the beam, the bending moment is $3 \times 4 - 9 \times 0.5 = +7.5$ kN m.

Example

A uniform cantilever of length 3.0 m (Figure 2.12) has a weight per metre of 120 kN. Determine the shear force and bending moment at distances of 1.0 m and 3.0 m from the free end if no other loads are carried by the beam.

At 1.0 m from the free end, there is 1.0 m of beam to the right and it has a weight of 120 kN (Figure 2.13(a)). Thus the shear force is +120 kN; it is positive because the forces are clockwise. The weight of this section can be considered to act at its centre of gravity which, because the beam is uniform, is at its midpoint. Thus the 120 kN weight force can be considered to be 0.5 m from the 1.0 m point and so the bending moment is $-120 \times 0.5 = -60$ kN m; it is negative because there is hogging.

At 3.0 m from the free end, there is 3.0 m of beam to the right and it has a weight of 360 kN (Figure 2.13(b)). Thus the shear force is +360 kN. The weight of this section can be considered to act at its midpoint, a distance of 1.5 m from the free end. Thus the bending moment is $-360 \times 1.5 = -540$ kN m.

120 kN/m

3.0 m

Figure 2.12   *Example*

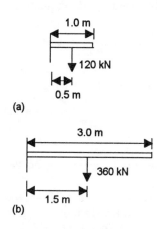

1.0 m

120 kN

0.5 m

(a)

3.0 m

360 kN

1.5 m

(b)

Figure 2.13   *Example*

Revision

1   A beam of length 4.0 m rests on supports at each end and a concentrated load of 500 N is applied at its midpoint. Determine the shear force and bending moment at distances of (a) 1.0 m, (b) 2.5 m from the right-hand end of the beam. Neglect the weight of the beam.

2   A cantilever has a length of 2 m and a concentrated load of 8 kN is applied to its free end. Determine the shear force and bending moment at distances of (a) 0.5 m, (b) 1.0 m from the fixed end. Neglect the weight of the beam.

3   A uniform cantilever of length 4.0 m has a weight per metre of 10 kN. Determine the shear force and bending moment at 2.0 m from the free end if no other loads are carried by the beam.

2.2.1 Shear force and bending moment diagrams

Figures which graphically show how the variations of the shear forces and bending moments along the length of a beam are termed *shear force diagrams* and *bending moment diagrams*. The two quantities are plotted above the centre line of the beam if positive and below it if negative. The following show such diagrams for commonly occurring situations.

1   *Simply supported beam with point load at mid-span*
Figure 2.14(a) shows the beam and the forces concerned, the weight of the beam being neglected. For a central load F, the reactions at each end will be $F/2$.

Consider the shear forces. At point A, the forces to the right are $F - F/2$ and so the shear force at A is $+F/2$; it is positive because the forces are clockwise about A. This shear force value will not change as we move along the beam from A until point C is reached. To the right of C we have just a force of $F/2$ and this gives a shear force of $-F/2$; it is negative because the forces are anticlockwise about it. To the left of C we have just a force of $F/2$ and this gives a shear force of $+F/2$; it is positive because the forces are clockwise about it. Thus at point C, the shear force takes on two values. For points between C and B, the forces to the left are constant at $F/2$ and so the shear force is constant at $-F/2$. Figure 2.14(b) shows the shear force diagram.

Consider the bending moments. At point A, the moments to the right are $F \times L/2 - F/2 \times L = 0$. The bending moment is thus 0. At point C the moment to the right is $F/2 \times L$ and so the bending moment is $+FL/2$; it is positive because sagging is occurring. At point B the moment to the right is zero, likewise that to the left $F \times L/2 - F/2 \times L = 0$. Between A and C the bending moment will vary, e.g. at one-quarter the way along the beam it is $FL/8$. In general, between A and C the bending moment a distance x from A is $Fx/2$ and between C and B is $Fx/2 - F(x - L/2) = F/2(L - x)$. Figure 2.14(c) shows the bending moment diagram. The maximum bending moment occurs under the load and is:

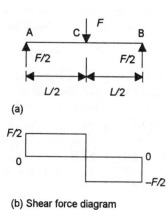

A   C   F   B

F/2   F/2

L/2   L/2

(a)

F/2

0   0

−F/2

(b) Shear force diagram

+FL/4

0

(c) Bending moment diagram

Figure 2.14   *Simply supported beam with point load*

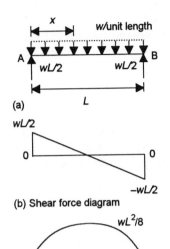

Figure 2.15 *Simply supported beam with distributed load*

(a)

(b) Shear force diagram

(c) Bending moment diagram

$$\text{maximum bending moment} = \frac{FL}{4} \qquad [1]$$

2   *Simple supported beam with uniformly distributed load*

Consider a simple supported beam which carries just a uniformly distributed load of w/unit length (Figure 2.15). The reactions at each end are $wL/2$.

Consider the shear force a distance x from the left-hand end of the beam. The load acting on the left-hand section of beam is wx. Thus the shear force is:

$$V = wL/2 - wx = w(\tfrac{1}{2}L - x) \qquad [2]$$

When $x = \tfrac{1}{2}L$, the shear force is zero. When $x < \tfrac{1}{2}L$ the shear force is positive and when $x > \tfrac{1}{2}L$ it is negative. Figure 2.15(b) shows the shear force diagram.

Consider the bending moment. At A the moment due to the beam to the right is $-wL \times L/2 + wL/2 \times L = 0$. At the midpoint of the beam the moment is $-wL/2 \times L/4 + wL/2 \times L/2 = wL^2/8$; the bending moment is thus $+wL^2/8$. At the quarter-point along the beam, the moment due to the beam to the right is $-3L/4 \times 3L/8 + wL/2 \times 3L/4 = 3wL^2/32$. In general, the bending moment due to the beam at distance x is:

$$M = -wx \times x/2 + wL/2 \times x = -wx^2/2 + wLx/2 \qquad [3]$$

Differentiating equation [3] gives $dM/dx = -wx + wL/2$. Thus $dM/dx = 0$ at $x = L/2$. The bending moment is thus a maximum at $x = L/2$, the value given by substituting this value in equation [3]:

$$\text{maximum bending moment} = \frac{wL^2}{8} \qquad [4]$$

Figure 2.15(c) shows the bending moment diagram.

3   *Cantilever with point load at free end*

Consider a cantilever which carries a point load F at its free end (Figure 2.16(a)), the weight of the beam being neglected. The shear force at any section will be $+F$, the shear force diagram thus being as shown in Figure 2.16(b). The bending moment at a distance x from the fixed end is:

$$M = -F(L - x) \qquad [5]$$

The minus sign is because the beam shows hogging. We have $dM/dx = F$ and thus the bending moment diagram is a line of constant slope F. At the fixed end, when $x = 0$, the bending moment is FL; at the free end it is 0.

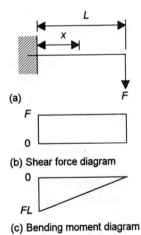

(a)

(b) Shear force diagram

(c) Bending moment diagram

Figure 2.16 *Cantilever with point load at free end*

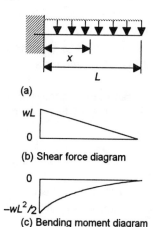

(a)

(b) Shear force diagram

(c) Bending moment diagram

Figure 2.17   *Cantilever with uniformly distributed load*

4   *Cantilever with uniformly distributed load*

Consider a cantilever which has just a uniformly distributed load of w per unit length (Figure 2.17(a)). The shear force a distance x from the fixed end is:

$$V = +w(L - x) \qquad [6]$$

Thus at the fixed end the shear force is $+wL$ and at the free end it is 0. Figure 1.23(b) shows the shear force diagram. The bending moment at a distance x from the fixed end is, for the beam to the right of the point, given by:

$$M = -w(L - x) \times (L - x)/2 = -\tfrac{1}{2}w(L - x)^2 \qquad [7]$$

This is a parabolic function. At the fixed end, where $x = 0$, the bending moment is $-\tfrac{1}{2}wL^2$. At the free end the bending moment is 0. Figure 2.17(c) shows the bending moment diagram.

In general, when drawing shear force and bending moment diagrams:

1   Between point loads, the shear force is constant and the bending moment gives a straight line.

2   Throughout a length of beam with a uniformly distributed load, the shear force varies linearly and the bending moment is parabolic.

3   The bending moment is a maximum when the shear force is zero. The proof of this follows.

4   The shear force is a maximum when the slope of the bending moment diagram is a maximum and zero when the slope is zero.

5   For point loads, the shear force changes abruptly at the point of application of the load by an amount equal to the size of the load.

As proof that the maximum value of the bending moment occurs at a point of zero shear force, consider a very small segment of beam (Figure 2.18) of length δx and which is supporting a uniformly distributed load of w per unit length. The load on the segment is $w\delta x$ and can be considered to act through its centre. The values of the shear force V and

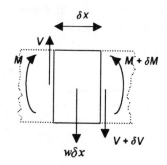

Figure 2.18   *Small segment of a beam*

bending moment M increase by δV and δM from one end of the segment to the other. If we take moments about the left-hand edge of the segment then:

$$M + V\delta x = w\delta x \times \delta x/2 + M + \delta M$$

Neglecting multiples of small quantities gives $V\delta x = \delta M$ and hence, as δx tends to infinitesimally small values, we can write:

$$V = \frac{dM}{dx} \qquad [8]$$

Thus $V = 0$ when $dM/dx = 0$.

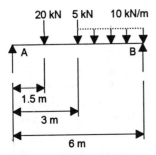

Example

A horizontal beam has a length of 6 m and is supported at its ends. A point load of 20 kN is applied at 1.5 m from the left-hand end and another point load of 5 kN at 3.0 m from the left-hand end. A uniformly distributed load of 10 kN/m is applied over the 3.0 m length between the 50 kN load and the right-hand end. Draw the shear force and bending moment diagrams and determine the position and size of the maximum bending moment and maximum shear force.

Figure 2.19   *Example*

Figure 2.19 described the arrangement. First the reactions at A and B are calculated. Taking moments about A:

$$6R_B = 20 \times 1.5 + 5 \times 3 + 30 \times 4.5$$

Hence $R_B = 30$ kN. For vertical equilibrium we have:

$$R_A + R_B = 20 + 5 + 30$$

and so $R_A = 25$ kN.

The shear force diagram will have constant values between the points where there are just point loads. Between A and the 20 kN point force, the shear force will have the value 25 kN. Between the 20 kN and 5 kN point forces the shear force will have the value $25 - 20 = 5$ kN. At the 5 kN load point, the shear force will drop to 0. Over the region where there is a distributed load the shear force will vary linearly from 0 to -30 kN. Figure 2.20(a) shows the shear force diagram.

For the region of the beam where we have just point loads, the bending moment will vary linearly between these points. It is zero at A, 37.5 kN m at the 20 kN point and 45 kN m at the 5 kN point. Over the distributed load portion of the beam the bending moment will vary parabolically to become 0 at B. Figure 2.20(b) shows the bending moment diagram.

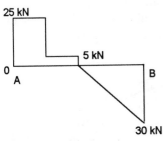

(a) Shear force diagram

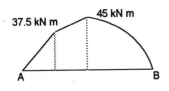

(b) Bending moment diagram

Figure 2.20   *Example*

The maximum bending moment is 45 kN m and occurs at the midpoint of the beam where the 5 kN point load is applied. The maximum shear force is 30 kN and occurs at B.

Revision

4 A beam of length 6 m is supported at points 1 m from each end. Draw the shear force and bending moment diagrams if it carries a uniformly distributed load of 10 kN per metre and determine the position and size of the maximum bending moment.

5 A beam of length 4 m is supported at points 1 m from each end. It carries point loads of 15 kN at one end, 10 kN at the other end and 80 kN at its midpoint. Draw the shear force and bending moment diagrams and determine the position and size of the maximum bending moment.

2.3 Bending stresses

When a beam bends, one surface becomes extended and so in tension and the other surface becomes compressed and so in compression (Figure 2.21). This implies that between the upper and lower surface there is a plane which is unchanged in length when the beam is bent. This plane is called the *neutral plane* and the line where the plane cuts the cross-section of the beam is the *neutral axis*.

Consider a beam, or part of a beam, where it can be assumed that it is bent to form the arc of a circle. This is termed *pure bending* and occurs when there is a constant bending moment, Figure 2.22 showing one way this can be realised. Consider the section through the beam aa which is a distance y from the neutral axis (Figure 2.23). It has increased in length as a consequence of the beam being bent. The strain it experiences will be its change in length ΔL divided by its initial unstrained length L. But for circular arcs, the arc length is the radius of the arc multiplied by the angle it subtends. Thus, since aa is of radius $R + y$, we have:

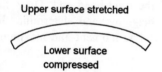

Upper surface stretched

Lower surface compressed

Figure 2.21   *Bending*

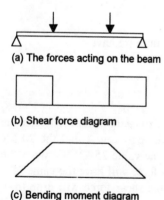

(a) The forces acting on the beam

(b) Shear force diagram

(c) Bending moment diagram

Figure 2.22   *Beam with central region in pure bending*

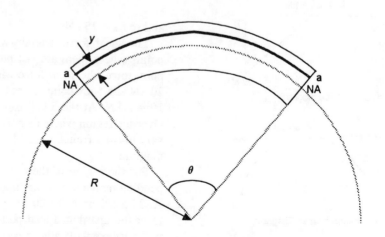

Figure 2.23   *Bending into the form of an arc of a circle*

$$L + \Delta L = (R + y)\theta$$

The neutral axis NA will, by definition, be unstrained. Thus:

$$L = R\theta$$

Hence, the strain on aa is:

$$\text{strain} = \frac{\Delta L}{L} = \frac{(R+y)\theta - R\theta}{R\theta}$$

and so:

$$\text{strain} = \frac{y}{R} \qquad\qquad [9]$$

The strain thus varies linearly through the thickness of the beam being larger the greater the distance y from the neutral axis. For a uniform rectangular cross-section beam the neutral axis is located symmetrically between the two surfaces and thus the maximum strain occurs on the surfaces of the beam.

Provided we can use Hooke's law the stress due to bending which is acting on aa is:

$$\text{stress} = E \times \text{strain} = \frac{Ey}{R} \qquad\qquad [10]$$

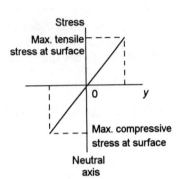

Figure 1.24 *Stress variation across beam section*

With a uniform rectangular cross-section beam, the maximum bending stresses will be on the surfaces. Figure 1.24 shows how the stress will vary across the section of the beam.

Example

A uniform square cross-section steel strip of side 4 mm is bent into a circular arc by bending it round a drum of radius 4 m. Determine the maximum strain and stress produced in the strip. Take the modulus of elasticity of the steel to be 210 GPa.

The neutral axis of the strip will be central and so the surfaces will be 2 mm from it. The radius of the neutral axis will be 4 + 0.002 m. Thus, using equation [9]:

$$\text{maximum strain} = \frac{y}{R} = \frac{2 \times 10^{-3}}{4.002} = 0.5 \times 10^{-3}$$

This will be the value of the compressive strain on the inner surface of the strip and the value of the tensile stress on the outer surface. The maximum stress will be:

maximum stress = $E \times$ max. strain

$$= 210 \times 10^9 \times 0.5 \times 10^{-3} = 105 \text{ MPa}$$

This will be tensile on the outer surface of the strip and compressive on the inner.

Revision

6   Steel strip is to be bent round a drum of radius 1 m. What is the maximum thickness of strip that can be bent in this way if the stress in the strip is not to exceed 100 MPa. The steel has a modulus of elasticity of 210 GPa.

2.3.1 The general bending equations ·

Consider a beam which has been bent into the arc of a circle so that its neutral axis has a radius R and an element of area dA in the cross-section of the beam at a distance y from the neutral axis (Figure 2.25). The element will be stretched as a result of the bending. The stress σ due to the bending acting on this element is given by equation [10] as Ey/R, where E is the modulus of elasticity of the material. The forces stretching this element aa are:

$$\text{force} = \text{stress} \times \text{area} = \sigma \delta A = \frac{Ey}{R} \delta A \qquad [11]$$

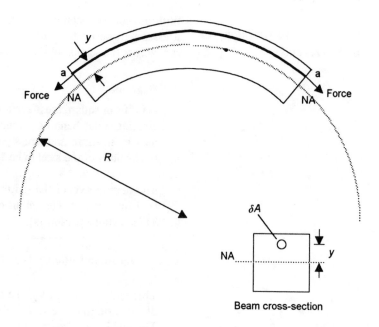

Figure 2.25   *Beam bent into the arc of a circle*

The moment of the force acting on this element about the neutral axis is:

$$\text{moment} = \text{force} \times y = \frac{Ey}{R}\delta A \times y = \frac{E}{R}y^2\delta A$$

The total moment M produced over the entire cross-section is the sum of all the moments produced by all the elements of area in the cross-section. Thus, if we consider each such element of area to be infinitesimally small, we can write:

$$M = \int \frac{E}{R}y^2 \, dA = \frac{E}{R}\int y^2 dA \qquad [12]$$

The integral is termed the *second moment of area I* of the section:

$$I = \int y^2 \, dA \qquad [13]$$

Thus equation [12] can be written as:

$$M = \frac{EI}{R} \qquad [14]$$

Since the stress σ on a layer a distance y from the neutral axis is yE/R then we can also write equation [14] as:

$$M = \frac{\sigma I}{y} \qquad [15]$$

Equations [14] and [15] are generally combined and written as the *general bending formula*:

$$\frac{M}{I} = \frac{\sigma}{y} = \frac{E}{R} \qquad [16]$$

This formula is only an exact solution for the case of pure bending, i.e. where the beam is bent into the arc of a circle and the bending moment is constant. However, many beam problems involve bending moments which vary along the beam. In these cases, equation [16] is generally still used since it provides answers which are usually accurate enough for engineering design purposes.

2.3.2 First moment of area

Consider the beam bent into the arc of a circle. The forces acting on a segment a distance y from the neutral axis is given by equation [11] as:

$$\text{force} = \text{stress} \times \text{area} = \sigma\delta A = \frac{Ey}{R}\delta A$$

The total longitudinal force will be the sum of all the forces acting on such segments and thus, when we consider infinitesimally small areas, is given by:

$$\text{total force} = \int \frac{Ey}{R}\, dA = \frac{E}{R} \int y\, dA$$

But the beam is only bent and so only acted on by a bending moment, there is no longitudinal force stretching the beam. Thus, since E and R are not zero, we must have:

$$\int y\, dA = 0 \qquad\qquad [17]$$

The integral $\int y\, dA$ is called the *first moment of area* of the section. The only axis about which we can take such a moment and obtain 0 is an axis through the centre of the area of the cross-section, i.e. the centroid of the beam. Thus the neutral axis must pass through the centroid of the section when the beam is subject to just bending.

Example

Determine the position of the neutral axis for the T-section beam shown in Figure 2.26.

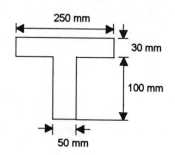

Figure 2.26   *Example*

The neutral axis will pass through the centroid. We can consider the T-section to be composed of two rectangular sections. The centroid of each will be at its centre. Hence, taking moments about the base of the T-section:

$$\text{total moment} = 250 \times 30 \times 115 + 100 \times 50 \times 50$$

$$= 1.11 \times 10^6 \text{ mm}^4$$

Hence the distance of the centroid from the base is (total moment)/(total area):

$$\text{distance from base} = \frac{1.11 \times 10^6}{250 \times 30 + 100 \times 50} = 89 \text{ mm}$$

Revision

7   Determine the position of the neutral axis from the base for the non-symmetrical I-section shown in Figure 2.27.

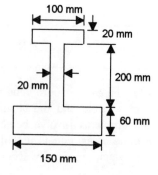

Figure 2.27   *Revision problem 7*

2.3.3 Second moments of area

The integral $\int y^2\, dA$ defines the *second moment of area I* about an axis. Consider a rectangular cross-section of breadth b and depth d (Figure 2.28). For a layer of thickness δy a distance y from the neutral axis, which passes through the centroid, the second moment of area for the layer is:

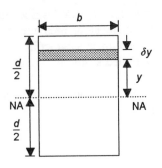

Figure 2.28 *Second moment of area*

second moment of area of strip $= y^2 \delta A = y^2 b \delta y$

The total second moment of area for the section is thus:

$$\text{second moment of area} = \int_{-d/2}^{d/2} y^2 b \, dy = \frac{bd^3}{12} \qquad [18]$$

If we had a second moment of area $I = \int y^2 \, dA$ of an area about an axis and then considered a situation where the area was moved by a distance h from the axis, the new second moment of area I_h would be:

$$I_h = \int (y+h)^2 \, dA = \int y^2 \, dA + 2h \int y \, dA + h^2 \int dA$$

But $\int y \, dA = 0$ and $\int dA = A$. Hence:

$$I_h = I + Ah^2 \qquad [19]$$

This is called the *theorem of parallel axes* and is used to determine the second moment of area about a parallel axis.

Figure 2.29 shows some values of second moments of area about the netural axes for commonly encountered beam sections.

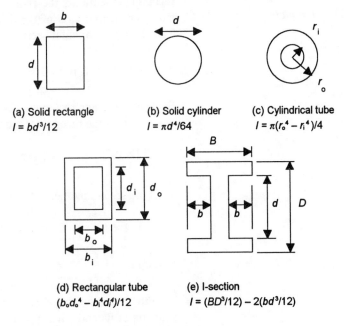

(a) Solid rectangle
$I = bd^3/12$

(b) Solid cylinder
$I = \pi d^4/64$

(c) Cylindrical tube
$I = \pi(r_o^4 - r_i^4)/4$

(d) Rectangular tube
$(b_o d_o^4 - b_i^4 d_i^4)/12$

(e) I-section
$I = (BD^3/12) - 2(bd^3/12)$

Figure 2.29 *Second moments of area*

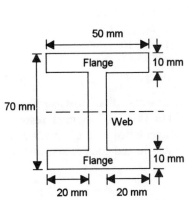

50 mm

Flange   10 mm

70 mm

Web

Flange   10 mm

20 mm   20 mm

Figure 2.30 *Example*

Example

Determine the second moment of area about the neutral axis of the I-section shown in Figure 2.30(a).

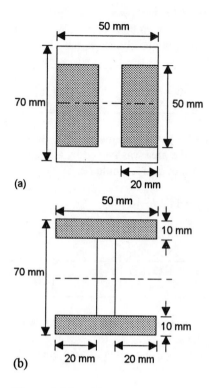

(a)

(b)

Figure 2.31   *Example*

One way of determining the second moment of area for such a section involves determining the second moment of area for the entire rectangle containing the section and then subtracting the second moments of area for the rectangular pieces 'missing' (Figure 2.31(a)).

Thus for the rectangle containing the entire section, the second moment of area is given by $I = bd^3/12$ as:

$$\frac{50 \times 70^3}{12} = 1.43 \times 10^6 \text{ mm}^4$$

For the two 'missing' rectangles, each will have a second moment of area of:

$$\frac{20 \times 50^3}{12} = 0.21 \times 10^6 \text{ mm}^4$$

Thus the second moment of area of the I-section is:

$$1.43 \times 10^6 - 2 \times 0.21 \times 10^6 = 1.01 \times 10^6 \text{ mm}^4$$

Another way of determining the second moment of area of the I-section is to consider it as three rectangular sections, one being the central rectangular section, the web, and the others a pair of rectangular sections, the flanges, with their neutral axes displaced from the neutral axis of the I-section by 30 mm (Figure 2.31(b)). The central rectangular section has a second moment of area of:

$$\frac{10 \times 50^3}{12} = 0.104 \times 10^6 \text{ mm}^4$$

Each of the outer rectangular areas will have a second moment of area given by the theorem of parallel axes as:

$$\frac{50 \times 10^3}{12} + 50 \times 10 \times 30^2 = 0.454 \times 10^6 \text{ mm}$$

Thus the second moment of area of the I-section is:

$$0.104 \times 10^6 + 2 \times 0.454 \times 10^6 = 1.01 \times 10^6 \text{ mm}^4$$

Example

A horizontal beam with a uniform rectangular cross-section of breadth 100 mm and depth 150 mm is 4 m long and rests on supports at its ends. It supports a concentrated load of 10 kN at its midpoint. Determine the maximum tensile and compressive stresses in the beam.

The second moment of area is:

$$I = bd^3/12 = 0.100 \times 0.150^3/12 = 2.8 \times 10^{-5} \text{ m}^4$$

The reactions at each support will be 5 kN and so the maximum bending moment, which will occur at the midpoint, is 10 kN m. The maximum bending stress will occur at the cross-section where the bending moment is a maximum and on the outer surfaces of the beam, i.e. $y = \pm 75$ mm. Using equation [22]:

$$\sigma = \frac{My}{I} = \pm \frac{10 \times 10^3 \times 0.075}{2.8 \times 10^{-5}} = \pm 26.8 \text{ MPa}$$

Revision

8   Determine the second moment of area of an I-section, about its horizontal neutral axis when the web is vertical, if it has rectangular flanges each 120 mm by 10 mm, a web of thickness 12 mm and an overall depth of 150 mm.

9   For the I-section in the above problem, determine the maximum bending moment that can be applied if the maximum bending stress is not to exceed 80 MPa.

2.3.4 Section modulus

For a beam which has been bent, the maximum stress σ_{max} will occur at the maximum distance y_{max} from the neutral axis. Thus, using equation [15], we can write:

$$M = \frac{I}{y_{max}} \sigma_{max}$$

The quantity I/y_{max} is a purely geometric function and is termed the *section modulus Z*. Thus:

$$M = Z\sigma_{max} \qquad [20]$$

with:

$$Z = \frac{I}{y_{max}} \qquad [21]$$

For a rectangular cross-section beam, the second moment of area $I = bd^3/12$ and the maximum stress occurs at the surfaces which are $d/2$ from the neutral axis. Thus $Z = (bd^3/12)/(d/2) = bd^2/6$. Standard section handbooks give values of section modulus for different section beams.

Example

A beam has a section modulus of 2×10^6 mm^3, what will be the maximum bending moment that can be used if the stress must not exceed 6 MPa?

Using equation [20]:

$$M = Z\sigma_{max} = 2 \times 10^6 \times 10^{-9} \times 6 \times 10^6 = 12 \text{ kN m}$$

Example

An I-section beam has a section modulus of 25×10^{-5} m³. What will be the maximum bending stress produced when the beam is subject to a bending moment of 30 kN m?

Using equation [20]:

$$\sigma_{max} = \frac{M}{Z} = \frac{30 \times 10^3}{25 \times 10^{-5}} = 120 \text{ MPa}$$

Example

A rectangular cross-section timber beam of length 4 m rests on supports at each end and carries a uniformly distributed load of 10 kN/m. If the stress must not exceed 8 MPa, what will be a suitable depth for the beam if its width is to be 100 mm?

For a simply supported beam with a uniform distributed load over its full length, the maximum bending moment is $wL^2/8$ (equation [4]) and thus the maximum bending moment for this beam is $10 \times 4^2/8 = 20$ kN m. Using equation [14]:

$$Z = \frac{M}{\sigma_{max}} = \frac{20 \times 10^3}{8 \times 10^6} = 2.5 \times 10^{-3} \text{ m}^3$$

For a rectangular cross-section $Z = bd^2/6$ and thus:

$$d = \sqrt{\frac{6Z}{b}} = \sqrt{\frac{6 \times 2.5 \times 10^{-3}}{0.100}} = 0.387$$

A suitable beam might thus be one with a depth of 400 mm.

Revision

10 A steel scaffold tube has a section modulus of 7.2×10^{-6} m³. What will be the maximum allowable bending moment on the tube if the bending stresses must not exceed 100 MPa?

11 An I-section beam has a section modulus of 3.2×10^{-5} m³. What will be the maximum allowable bending moment on the beam if the bending stresses must not exceed 150 MPa?

12 Determine the section modulus required of a steel beam which is to span a gap of 6 m between two supports and support a uniformly distributed load over its entire length, the total distributed load amounting to 65 kN. The maximum bending stress permissible is 165 MPa.

Problems   1   A beam of length 4.0 m is supported at its ends and carries a concentrated load of 20 kN at its midpoint. Determine the shear force and bending moment at distances of (a) 0.5 m and (b) 1.0 m from the right-hand end. Neglect the weight of the beam.

2   A uniform beam of length 4.0 m is supported at its ends and has a weight of 10 kN/m. It carries no other loads. Determine the shear force and bending moment at distances of (a) 0.5 m and (b) 1.0 m from the right-hand end.

3   A beam of length 6 m is supported at both ends and carries a point load of 40 kN at its midpoint. Draw the shear force and bending moment diagrams and determine the position and value of the maximum bending moment.

4   A beam of length 6 m is supported at both ends and carries a point load of 60 kN at a distance of 2 m from one end. Draw the shear force and bending moment diagrams and determine the position and value of the maximum bending moment.

5   A beam of length 4 m is supported at points 1 m and 3 m from one end. It carries point loads of 20 kN at each end. Determine the maximum shear force and the maximum bending moment.

6   A cantilever of length 4 m carries point loads of 30 kN at 1 m from the fixed end, 20 kN at 3 m from the fixed and 10 kN at the free end. Draw the shear force and bending moment diagrams and determine the positions and values of the maximum shear stress and bending moment.

7   A beam of length 10 m is supported at one end and at a point 8 m from the supported end. It carries a uniformly distributed load of 16 kN/m over its entire length and a point load of 40 kN at the unsupported end. Draw the shear force and bending moment diagrams and determine the positions and values of the maximum shear stress and bending moment.

8   A steel strip of thickness 0.8 mm is bent round a pulley of radius 200 mm. What is the maximum stress produced in the wire as a result of the bending? The steel has a modulus of elasticity of 210 GPa.

9   Determine the position of the neutral axis of a T-section if the top of the T is a rectangle 100 mm by 10 mm and the stem of the T is a rectangle 120 mm by 10 mm.

10   Determine the second moment of area of a rectangular section of breadth 50 mm and depth 100 mm.

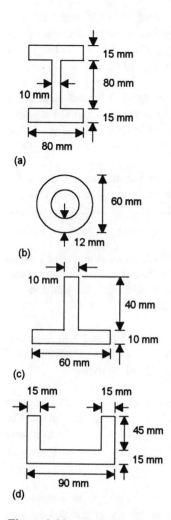

15 mm

80 mm

10 mm

15 mm

80 mm

(a)

60 mm

12 mm

(b)

10 mm

40 mm

10 mm

60 mm

(c)

15 mm   15 mm

45 mm

15 mm

90 mm

(d)

Figure 2.32   *Problem 15*

11 Determine the position of the neutral axis of a T-section if the top of the T is a rectangle 150 mm by 10 mm and the stem of the T is a rectangle 90 mm by 10 mm.

12 A beam has a rectangular cross section of width 60 mm and depth 100 mm. Determine the maximum bending moment that can be applied if the bending stresses are not to exceed ±150 MPa.

13 A rectangular tube section has an overall width of 80 mm and a depth of 150 mm. The inner walls have a width of 60 mm and a depth of 130 mm, the walls of the tube being 10 mm thick. Determine the maximum tensile and compressive stresses such a section will experience when subjected to a bending moment of 20 kN m.

14 A beam has to support loading which results in a maximum bending moment of 25 kN m. If the maximum permissible bending stress is 7 MPa, what will be the required section modulus?

15 For the beam sections shown in Figure 2.32, determine the position of the neutral axis, the second moment of area about the neutral axis and the section modulus for the top edge.

16 A uniform beam of length 6 m and section modulus 2.3×10^{-3} m^3 is supported at its ends. Point loads of 40 kN are carried at 1.5 m from each end. What will be the maximum stress experienced by the beam?

3 Torsion

3.1 Introduction

As discussed in Chapter 1 in relation to the loading of structures, one basic form of loading is torsion. Thus for a motor being used to rotate a load (Figure 3.1), the shaft becomes twisted as a result of the motor rotating its end of the shaft. It twists until the resisting torque offered by the shaft material balances the torque to be transmitted. The twisted shaft is said to be in *torsion*. *Torque T* is defined as the turning moment of an applied force about an axis; for a force F with a radius of rotation r about the axis (Figure 3.2):

$$\text{torque } T = Fr \qquad [1]$$

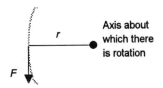

Figure 3.1   *Motor rotating a load*

This chapter is a discussion of the twisting of shafts, the stresses produced and the power that can be transmitted by shafts. Both solid and hollow circular section shafts are considered.

3.1.1 Shear stress and strain

In discussions of torsion we will be concerned with shear stress and shear strain. The following is a review of such terms. *Shear* is said to occur if the forces applied to a block of material result in a tendency for one face of the material to slide relative to the opposite face (Figure 3.3). With shear the area over which forces act is in the same plane as the line of action of the forces, unlike direct stress where the area is at right angles to the forces. The force per unit area is called the *shear stress* τ:

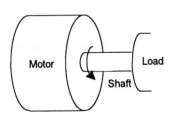

Figure 3.2   Torque $= Fr$

$$\text{shear stress } \tau = \frac{F}{A} \qquad [2]$$

The unit of shear stress is the pascal (Pa) when the force is in newtons and the area in square metres. The deformation produced by the shear is, for Figure 3.3, XY relative to a parallel face a distance L away. The *shear strain* is this distance XY divided by L. But tan ϕ = XY/L and for the small angles generally involved, tan ϕ is approximately ϕ. Thus shear strain is expressed as the angular deformation ϕ:

$$\text{shear strain} = \phi \qquad [3]$$

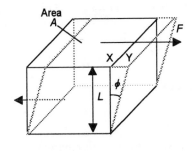

Figure 3.3   *Shear*

The unit used for ϕ is the radian and, since the radian is a ratio, shear strain can be either expressed in units of radians or as a ratio and so without units.

Shear stress is proportional to shear strain for many materials and the stress/strain ratio is called the *shear modulus G* or *modulus of rigidity*:

$$\text{shear modulus} = \frac{\text{shear stress}}{\text{shear strain}} \qquad [4]$$

The unit of the shear modulus is the same as that of shear stress, since shear strain has no units, and is thus Pa.

As illustrations of shear situations, Figure 3.4(a) shows shear occurring in a riveted joint, the form of joint shown being termed a lap joint. The rivet is in shear as a result of the forces applied to the plates joined by the rivet. Figure 3.4(b) shows shear stresses being applied by a punch to deform a sheet of material and give a cupped indentation or to punch a circular hole in the material.

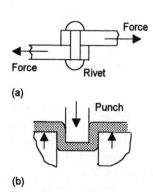

(a)

(b)

Figure 3.4   *Examples of shear*

Example

What forces are required to shear a lap joint made using a single 25 mm diameter rivet if the maximum shear stress the rivet can withstand is 250 MPa?

The joint is as shown in Figure 3.4(a). Using equation [2]:

force = shear stress × area

$$= 250 \times 10^6 \times \tfrac{1}{4}\pi \times 0.025^2 = 123 \text{ kN}$$

Example

What is the maximum load that can be applied to the pin coupling shown in Figure 3.5 if the pin has a diameter of 10 mm and the shear stress in the pin is not to exceed 50 MPa.

There are two surfaces in the pin being sheared by the action of the forces, thus the area to be sheared is double the cross-sectional area of the pin. Using equation [2]:

force = shear stress × area

$$= 50 \times 10^6 \times 2 \times \tfrac{1}{4}\pi \times 0.010^2 = 7.85 \text{ kN}$$

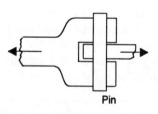

Figure 3.5   *Example*

Revision

1   What is the minimum diameter required for the bolt in Figure 3.6 if the shear stress in it is not to exceed 90 MPa when the forces applied to the components joined by the bolt are 30 kN?

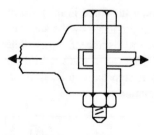

Figure 3.6   *Revision problem 1*

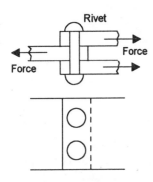

Figure 3.7 *Revision problem 3*

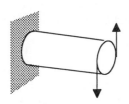

(a) Before torque applied

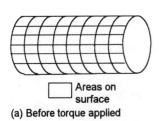

(b) With torque applied

Figure 3.9 *Circular shaft: (a) before, (b) after torque applied*

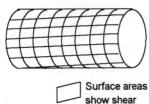

2   What is the maximum diameter hole that can be punched in an aluminium plate of thickness 14 mm if the punching force is limited to 50 kN? The shear strength, i.e. the maximum shear stress the material can withstand before failure, is 90 MPa.

3   Three steel plates are joined by two rivets (Figure 3.7), with each rivet having a diameter of 15 mm. What is the maximum force that can be applied if the shear stress in the rivets is not to exceed 200 MPa?

4   A metal cube has a side of 20 mm. Opposite faces have forces applied to them to give a shear strain of 0.0005. What is the relative displacement of the opposite faces?

5   The shear modulus for a material is 80 GPa. What will be the shear strain when the shear stress is 20 MPa?

3.2 Torsion of circular shafts

Torsion is the term used for the twisting of a structural member when it is acted on by torques (Figure 3.8) so that rotation is produced about the longitudinal axis of one end of the member with respect to the other. For simplification in deriving equations for torsion we will make the following assumptions:

1   The shaft has a uniform circular cross-section.
2   The shaft material is uniform throughout and shear stress is proportional to shear strain.
3   The shaft is straight and initially unstressed.
4   The axis of the twisting moment is the axis of the shaft.
5   Plane transverse sections remain plane after twisting. As a result, each circular section is rotated different amounts and results in shear forces (Figure 3.9).

Consider such a shaft of radius r and length L. If the angle of twist is θ then the situation is as shown in Figure 3.10; AC is the initial position of a line along the surface of the bar and BC is its new position as a result of the end of the bar being rotated through the arc AB. The arc AB subtends an angle θ and so arc AB = $r\theta$. But AB also equals $L\phi$, where ϕ is the resulting shear strain; if you think of a strip on the surface of the shaft it becomes sheared as shown in Figure 3.11.

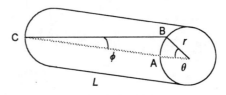

Figure 3.10 *Twisting of a cylindrical shaft*

Figure 3.8 *Torsion*

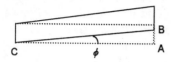

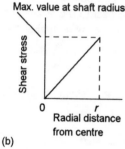

Figure 3.11   *Shear of strip on the surface*

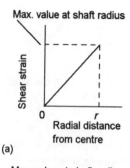

Figure 3.12   *Shear strain and shear stress*

Thus:

$$\phi L = r\theta$$

and so the shear strain is:

$$\phi = \frac{r\theta}{L} \tag{5}$$

This states that for a given angle of twist per unit length, i.e. θ/L, the shear strain is proportional to the distance r from the central axis (Figure 3.12(a)). The shear stress τ at this radius is thus, using equation [4]:

$$\tau = \frac{Gr\theta}{L} \tag{6}$$

where G is the shear modulus. The shear stress is, for a particular angle of twist per unit length and material, proportional to the distance r from the central axis (Figure 3.12(b)).

As a simplification, consider the torsion of a thin-walled tube element of the shaft with a radius r and thickness Δr (Figure 3.13). The tube is considered to be thin enough for the assumption to be made that the shear stress is uniform throughout the thickness of the tube wall. Thus for the element of area ΔA the shear force is:

$$\text{shear force} = \text{area} \times \text{shear stress} = \Delta A \times \tau$$

where τ is the shear stress acting on the element. The shear stress acts circumferentially to the tube across the area of the cross-section. Thus the shear force on the total area of the tube wall is:

$$\text{shear force} = 2\pi r \Delta r \times \tau$$

This shear force is acting at a radius r in a tangential direction. It gives a torque T about the axis of the tube of:

$$T = \text{shear force} \times \text{radius} = 2\pi r \Delta r \times \tau \times r = 2\pi r^2 \Delta r \times \tau \tag{7}$$

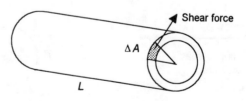

Figure 3.13   *Shear force acting on an element of a thin-walled tube of thickness Δt and radius r*

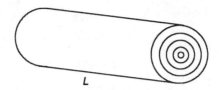

Figure 3.14   *Solid tube as composed of a large number of thin-walled tubes of different radii*

Using equation [6] we have $\tau = Gr\theta/L$ and so we can write equation [7] as:

$$T = 2\pi r^2 \Delta r \times \frac{Gr\theta}{L} = \frac{G\theta}{L} \times 2\pi r^3 \Delta r \qquad [8]$$

Equation [8] gives the shear force for a thin-walled tube. We can think of a solid rod as being composed of a large number of such tubes (Figure 3.14) and thus the shear force is the sum of all the contributions from each of the tubes which go to make up the solid tube. Thus, when we make the tubes infinitesimally thin:

$$T = \int_0^r \frac{G\theta}{L} \times 2\pi r^3 \, dr = \frac{G\theta}{L} \int_0^r 2\pi r^3 \, dr \qquad [9]$$

This is written as:

$$T = \frac{G\theta J}{L} \qquad [10]$$

where J is the *polar second moment of area* about the shaft axis and defined by:

$$\text{polar second moment of area } J = \int_0^r 2\pi r^3 \, dr \qquad [11]$$

Equations [6] and [10] are often written in the following form, it being referred to as the *general equation for the torsion of circular cross-section shafts*:

$$\frac{T}{J} = \frac{\tau}{r} = \frac{G\theta}{L} \qquad [12]$$

3.2.1 Polar second moment of area

The polar second moment of area J of a shaft about its axis is defined by equation [11]. Thus, with a solid shaft of diameter D:

$$J = 2\pi \int_0^{D/2} r^3 \, dr = \frac{\pi D^4}{32} \qquad [13]$$

For a hollow shaft with external diameter D and internal diameter d:

$$J = 2\pi \int_{d/2}^{D/2} r^3 \, dr = \frac{\pi}{32}(D^4 - d^4) \qquad [14]$$

The units of J are m⁴ when the diameters are in metres.

Example

What is the minimum diameter of a solid shaft if it is to transmit a torque of 30 kN and the shear stress in the shaft is not to exceed 80 MPa?

For a solid shaft the polar second moment of area about its axis is given by equation [13] as $J = \pi D^4/32$ and thus equation [12] gives:

$$T = \frac{J\tau}{r} = \frac{2J\tau}{D} = \frac{\pi D^3 \tau}{16}$$

and so:

$$D^3 = \frac{16T}{\pi\tau} = \frac{16 \times 30 \times 10^3}{\pi \times 80 \times 10^6}$$

Thus the minimum diameter shaft is 0.124 m.

Example

Sketch a graph of the shear stress across a section of a 50 mm diameter solid shaft when subject to a torque of 200 N m.

For a solid shaft the polar second moment of area about its axis is given by equation [13] as $J = \pi D^4/32 = \pi \times 0.050^4/32 = 6.14 \times 10^{-7}$ m⁴. Thus equation [12] gives:

$$\tau = \frac{Tr}{J} = \frac{200r}{6.14 \times 10^{-7}} = 3.26 \times 10^8 r \text{ Pa}$$

When $r = 0$ then the shear stress is 0; when $r = 25$ mm then the shear stress is 8.15 MPa. Since the shear stress is proportional to the distance r from the axis, the graph is as shown in Figure 3.15.

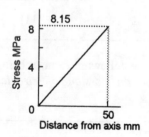

Figure 3.15   *Example*

Example

A steel tube of length 4 m has an external diameter of 10 mm and an internal diameter of 6 mm. Determine the maximum and minimum shear stresses in the tube when, with one end fixed, the other is rotated through 30°. The steel has a shear modulus of 80 GPa.

Using equation [12], with $\theta = 30° = \pi/6$ radians:

$$\tau = \frac{G\theta r}{L} = \frac{(80 \times 10^9 \times (\pi/6)r}{4} = 1.05 \times 10^{10} r \text{ Pa}$$

The maximum shear stress occurs at the maximum radius of 5 mm and thus is 52.5 MPa and the minimum shear stress occurs at the minimum radius of 3 mm and thus is 31.5 MPa.

Revision

6 A solid steel shaft has a diameter of 60 mm. Determine the maximum torque that can be applied to the shaft if the maximum permissible shear stress is 40 MPa.

7 A solid steel shaft has a diameter of 60 mm. Determine the maximum torque that can be applied to the shaft if the maximum permissible twist per unit length is 1° per metre. The steel has a shear modulus of 80 GPa.

8 Determine the external diameter of a tube needed to transmit torque of 30 kN m if it has an external diameter which is twice its internal diameter and the shear stress is not to exceed 80 MPa.

9 A tubular shaft has an internal diameter which is half its external diameter and is subject to a torque of 50 kN m. What external diameter will be required if the shear stress in the shaft is not to exceed 80 MPa?

3.2.2 Polar section modulus

For a shaft of radius R, equation [13] gives for the relationship between the torque T and the maximum shear stress τ_{max}:

$$\frac{T}{J} = \frac{\tau_{max}}{R}$$

and so:

$$T = \frac{J}{R}\tau_{max} = Z_p \tau_{max} \qquad [15]$$

The *polar section modulus of section* Z_p is defined as:

$$Z_p = \frac{J}{R} \qquad [16]$$

For a solid shaft where $J = \pi D^4/32$, with $D = 2R$, then $Z_p = \pi D^3/16$.

Example

Determine the polar section modulus for a uniform solid shaft of diameter 40 mm.

For a solid shaft where $J = \pi D^4/32$, with $D = 2R$, then $Z_p = J/R = \pi D^3/16 = \pi \times 0.040^3/16 = 1.26 \times 10^{-5}$ m^3.

Revision

10   A shaft has a polar section modulus of 1.5×10^{-4} m^3 and is subject to a torque of 5 kN m. What will be the maximum shear stress in the shaft?

3.3 Transmission of power

For a shaft of radius r, the distance travelled by a point on its surface when it rotates through one revolution is $2\pi r$. When it rotates through n revolutions the distance travelled is $2\pi rn$. If is rotating at n revolutions per second then the distance travelled per second is $2\pi rn$. If the shaft is being rotated by a torque T, the force acting at a radius r is $F = T/r$. The work done per second will be the product of the force and the distance moved per second by the point to which the force is applied. Thus:

work done per second = $F \times 2\pi rn = 2\pi nT$

But the work done per second is the power. Thus the power transmitted per second is:

$$\text{power} = 2\pi nT \qquad [17]$$

In one revolution the angle rotated by the point on the shaft surface is 2π radians. If n revolutions are completed per second then the angle rotated in one second is $2\pi n$. The angular velocity ω is thus $2\pi n$. Thus:

$$\text{power transmitted} = \omega T \qquad [18]$$

With the angular velocity in rad/s and the torque in N m, the power is in watts.

We can thus use equation [18] to determine the power that can be transmitted by an electric motor rotating its shaft. However, in the case of an internal combustion engine, the torque transmitted through the output shaft is not constant but varies during the cycle (Figure 3.16). The power that can be transmitted in this case is thus calculated from the mean torque during the cycle. But when we need to calculate the diameter required for the shaft in order that the maximum shear stress should not be exceeded, the peak value of the torque has to be used.

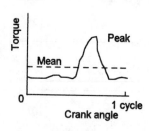

Figure 3.16   *Torque delivered by an internal combusion engine*

Example

Determine the power that can be transmitted by a solid steel shaft of diameter 100 mm which is rotating at 5 rev/s if the shear stress in the shaft is not to exceed 70 MPa.

For a solid shaft, $J = \pi D^4/32 = \pi \times 0.100^4/32 = 9.82 \times 10^{-6}$ m^4. The maximum torque that can act is given by equation [13] as $J\tau/R = 2J\tau/D = 2 \times 9.82 \times 10^{-6} \times 70 \times 10^6/0.100 = 13.7$ kN m. Thus, using equation [17]:

$$\text{max. power} = 2\pi nT = 2\pi \times 5 \times 13.7 \times 10^3 = 430 \text{ kW}$$

Example

The drive shaft of a car is a tube with an external diameter of 50 mm and an internal diameter of 47 mm. Determine the maximum shear stress in the shaft when it is transmitting a power of 70 kW and rotating at 80 rev/s.

The maximum torque is given by equation [17] as:

$$\text{torque } T = \frac{P}{2\pi n} = \frac{70 \times 10^3}{2\pi \times 80} = 139 \text{ N m}$$

For the tube, $J = \pi(D^4 - d^4)/32 = \pi(0.050^4 - 0.047^4)/32 = 1.35 \times 10^{-7}$ m^4. The maximum shear stress is given by equation [13] as

$$\text{max. shear stress} = \frac{Tr}{J} = \frac{139 \times 0.025}{1.35 \times 10^{-7}} = 25.7 \text{ MPa}$$

Revision

11 A solid shaft of diameter 80 mm rotates at 5 rev/s and transmits a power of 50 kW. Determine the maximum shear stress in the shaft.

12 A tubular shaft has an external diameter of 150 mm and an internal diameter of 100 mm. Determine the maximum power that can be transmitted by the shaft when rotating at 3 rev/s if the shear stress in the shaft is not to exceed 50 MPa.

Problems   1 Two plates lap each other and are riveted together by ten rivets, each of diameter 8 mm. If the shear stress in a rivet is not to exceed 5 MPa, what is the maximum shear force which the joint can withstand?

2 A metal cube has a side of 30 mm. Opposite faces have forces applied to them to give a shear strain of 0.0002. What is the relative displacement of the opposite faces?

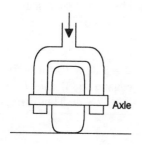

Axle

Figure 3.17   *Problem 4*

3   A punch of diameter 35 mm is used to punch a hole through a metal sheet of thickness 4 mm. What force will be required if the shear stress needed to shear the material is 50 MPa?

4   What is the maximum shear stress in the wheel axle shown in Figure 3.17 if the axle has a diameter of 40 mm and the load acting on the wheel is 50 kN?

5   Determine the minimum diameter a solid shaft can have if it is to transmit a torque of 1.2 kN m and the shear stress in the shaft is not to exceed 50 MPa.

6   Determine the maximum torque that can be transmitted by a solid shaft of diameter 40 mm if the shear stress is not to exceed 100 MPa.

7   A tubular shaft has an external diameter of 50 mm and an internal diameter of 25 mm. When subject to a torque of 1.5 kN m it is twisted through 1° per metre. Determine (a) the maximum shear stress in the shaft, (b) the shear modulus of the material, (c) the maximum shear strain.

8   A tubular shaft and a solid shaft are made from the same material, have the same length and the same external diameter and are subject to the same torque. The internal diameter of the tubular shaft is 0.6 times its external diameter. How do (a) the maximum shear stresses, (b) the angles of rotation, (c) the masses compare?

9   A tubular drive shaft has an external diameter of 160 mm and an internal diameter of 80 mm. What will be the maximum shear stress in the shaft when it is subject to a torque of 40 kN m?

10   Compare the torque that can be transmitted by a tubular shaft with that of a solid shaft if they both have the same material, mass, length and allowable shear stress.

11   A tubular drive shaft has an external diameter of 90 mm and an internal diameter of 64 mm. Determine the maximum and minimum shear stresses in the shaft when it is subject to a torque of 5 kN m.

12   Determine the maximum shear stress produced in a 6 mm diameter bolt when it is tightened by a spanner which applies a torque of 7 N m.

13   A shaft has a polar section modulus of 2.0×10^{-4} m^3. What will be the maximum torque that can be used with the shaft if the shear stress in the shaft is not to exceed 50 MPa?

14   A tubular shaft has an inner diameter of 30 mm and an external diameter of 42 mm and has to transmit a power of 60 kW. What will

be the limiting frequency of rotation of the shaft if the shear stress in the shaft is not to exceed 50 MPa?

15 The motor of an electric fan delivers 150 W to rotate the fan blades at 18 rev/s. Determine the smallest diameter of solid shaft that can be used if the shear stress in the shaft is not to exceed 80 MPa.

16 A solid shaft has a diameter of 30 mm and is used to transmit a power of 25 kW at 1000 rev/min. Determine (a) the maximum shear stress in the shaft and (b) the angle of twist over a length of 2 m. The material has a shear modulus of 80 GPa.

17 A tubular steel shaft has an external diameter of 50 mm and a wall thickness of 6.5 mm. Determine the torque that can be transmitted by the shaft if the shear stress in the shaft is not to exceed 15 MPa and the power that can be transmitted when the shaft rotates at 50 rev/s.

18 Two solid circular steel shafts of the same diameter of 80 mm are connected by a flanged coupling having six bolts each of 20 mm diameter and on a pitch circle of diameter 140 mm (Figure 3.18). The shear stress in a bolt must not exceed 70 MPa and that in the shafts must not exceed 120 MPa. What is the maximum power that can be transmitted at 5 rev/s? Note that there are two limiting conditions, that due to shear stress in the bolts and that due to shear stress in the shaft and it is necessary to determine which one determines the maximum torque and hence maximum power.

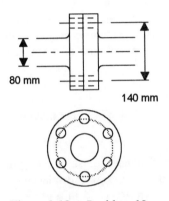

80 mm

140 mm

Figure 3.18   *Problem 18*

4 Linear and angular motion

4.1 Introduction

This chapter is concerned with the behaviour of dynamic mechanical systems when there is uniform acceleration. The terms and basic equations associated with linear motion with uniform acceleration and angular motion with uniform angular acceleration, Newton's laws of motion, moment of inertia, linear and angular kinetic energy and the effects of friction are revised and applied to the solution of mechanical system problems.

4.2 Linear motion

The following are basic terms used in the description of linear motion:

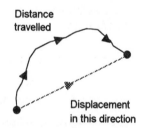

Figure 4.1   *Distance and displacement*

1   *Distance and displacement*

The term distance tends to be used for distances measured along the path of an object, whatever form the path takes; displacement, however, tends to be used for the distance travelled in a particular straight line direction (Figure 4.1). For example, if an object moves in a circular path the distance travelled is the circumference of the path whereas the displacement might be zero if it ends up at the same point it started from. Distance is thus a scalar quantity possessing only magnitude whereas displacement is a vector quantity with both magnitude and direction.

For motion in a straight line, the displacement in the direction of the line is numerically the same as the distance along that line.

2   *Speed and velocity*

Speed is the rate a distance s varies with time t and is a scalar quantity since the speed of a body is independent of the direction of motion. On a graph of distance against time, speed is the slope of the graph (Figure 4.2).

Velocity in a particular direction is the rate at which displacement s in that direction changes with time t and is a vector quantity since the velocity of a body is dependent on the direction of motion. On a graph of displacement, in a particular direction, against time, velocity is the slope of the graph (Figure 4.3).

For motion in a straight line, the velocity in the direction of the line is numerically the same as the speed along that line.

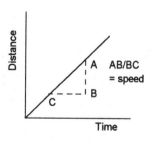

Figure 4.2   *Speed*

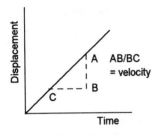

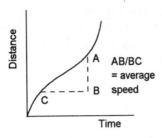

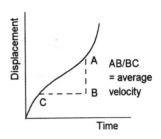

Figure 4.3   *Velocity*

Figure 4.4   *Average speed*

Figure 4.5   *Average velocity*

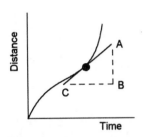

Figure 4.6   *Instantaneous speed = AB/BC*

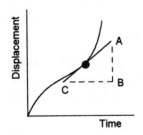

Figure 4.7   *Instantaneous velocity = AB/BC*

3   *Average speed and average velocity*
The average speed over some time interval is the distance covered in that time interval divided by the duration of the time interval (Figure 4.4):

$$\text{average speed} = \frac{\text{distance travelled}}{\text{time taken}} = \frac{\Delta s}{\Delta t} \qquad [1]$$

where Δs is the change in distance s when the time t changes by Δt.

The average velocity in a particular direction over some time interval is the displacement in that direction divided by the duration of the time interval (Figure 4.5):

$$\text{average velocity} = \frac{\text{displacement occurring}}{\text{time taken}} = \frac{\Delta s}{\Delta t} \qquad [2]$$

where Δs is the change in displacement s in a particular direction when the time t changes by Δt.

4   *Instantaneous speed and instantaneous velocity*
The speed at an instant of time can be considered to be the average speed over a time interval when we make the time interval vanishingly small. It is thus the slope of a tangent to a graph of distance against time at a particular point on the graph (Figure 4.6). It can be expressed as:

$$\text{speed} = \lim_{\Delta t \to 0} \frac{\Delta s}{\Delta t} = \frac{ds}{dt} \qquad [4]$$

The velocity at an instant of time can be considered to be the average velocity over a time interval when we make the time interval vanishingly small. It is thus the slope of a tangent to a graph of displacement, in a particular direction, against time at a particular point on the graph (Figure 4.7). It can be expressed as:

$$\text{velocity } v = \lim_{\Delta t \to 0} \frac{\Delta s}{\Delta t} = \frac{ds}{dt} \qquad [5]$$

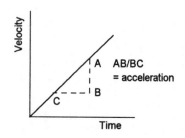

Figure 4.8   *Acceleration*

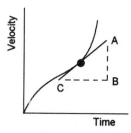

Figure 4.9   *Average acceleration*

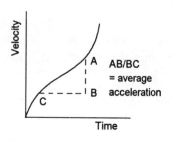

Figure 4.10   *Instantaneous acceleration = AB/BC*

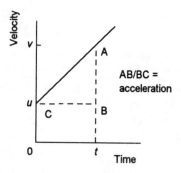

Figure 4.11   *Acceleration*

5   *Uniform speed and uniform velocity*
Uniform speed occurs when equal distances are covered in equal intervals of time, however small we consider the time intervals. Uniform velocity occurs when equal distances are being covered in the same straight line direction, however small we consider the time intervals.

6   *Acceleration*
Acceleration is the rate of change of velocity v with time t. On a graph of velocity against time, acceleration is the slope of the graph (Figure 4.8). Acceleration is a vector quantity. The term retardation is often used to describe a negative acceleration, i.e. when the object has a decreasing velocity. Note that acceleration can occur without any alteration in speed, e.g. an object moving in a circular path with uniform speed. There is an acceleration because the velocity is changing and this can result from a change in its magnitude, i.e. the speed, and/or a change in direction.

7   *Average acceleration*
Average acceleration is the change in velocity occurring over some interval divided by the duration of that time interval (Figure 4.9):

$$\text{average acceleration} = \frac{\text{change of velocity}}{\text{time taken for change}} = \frac{\Delta v}{\Delta t} \qquad [6]$$

where Δv is the change in velocity in a particular direction occurring in a time interval Δt.

8   *Instantaneous acceleration*
The acceleration at an instant of time can be considered to be the average acceleration over a time interval when we make the time interval vanishingly small (Figure 4.10):

$$\text{acceleration} = \lim_{\Delta t \to 0} \frac{\Delta v}{\Delta t} = \frac{dv}{dt} \qquad [7]$$

7   *Uniform acceleration*
Uniform acceleration occurs when the velocity changes by equal amounts in equal intervals of time, however small the time interval.

4.2.1 Motion with constant acceleration

Consider an object which is travelling in a straight line and accelerated uniformly from an initial velocity u to a final velocity v in a time t (Figure 4.11). The acceleration a is the change in velocity divided by the time interval concerned and so is:

$$a = \frac{v-u}{t}$$

Hence we can write:

$$v = u + at \qquad\qquad [8]$$

The average velocity over this time is $\frac{1}{2}(u + v)$ and thus the distance s travelled along the straight line path is:

$$s = \text{average velocity} \times \text{time} = \frac{1}{2}(u + v)t$$

Substituting for v by means of equation [8] gives:

$$s = \frac{1}{2}(u + u + at)t$$

and so:

$$s = ut + \frac{1}{2}at^2 \qquad\qquad [9]$$

We can arrive at equation [9] in a more general manner by using equation [5]. Since $v = ds/dt$ then, integrating both sides of the equation between the values at zero time and time t gives:

$$\int_0^s \frac{ds}{dt}\, dt = \int_0^t v\, dt$$

and so:

$$s = \int_0^t v\, dt \qquad\qquad [10]$$

This integral describes the area under the graph of velocity against time between zero time and time t. For motion with uniform acceleration, the graph will be of the form shown in Figure 4.12. The area is composed of two elements, that of a rectangle with area ut and that of a triangle with area $\frac{1}{2}(v - u)t$. Since $(v - u)/t$ is the acceleration a, the total area is:

$$s = ut + \frac{1}{2}at^2$$

We can derive a further equation for motion with uniform acceleration. If we square equation [8] we obtain:

$$v^2 = (u + at)^2 = u^2 + 2uat + a^2t^2 = u^2 + 2a(ut + \frac{1}{2}at^2)$$

Hence, using equation [9]:

$$v^2 = u^2 + 2as \qquad\qquad [11]$$

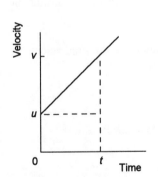

Figure 4.12 *Uniform acceleration*

Example

An object moves along a straight line from point A to point B with a uniform acceleration of 2 m/s^2. If the time taken is 10 s and the velocity at B is 25 m/s, determine the initial velocity at point A and the distance AB.

Using equation [8], i.e. $v = u + at$, then:

$$25 = u + 2 \times 10$$

Hence $u = 5$ m/s. Using equation [9], i.e. $s = ut + \frac{1}{2}at^2$, then:

$$s = 5 \times 10 + \frac{1}{2} \times 2 \times 10^2 = 150 \text{ m}$$

Alternatively we could have used equation [11], i.e. $v^2 = u^2 + 2as$, then:

$$25^2 = 5^2 + 2 \times 2 \times s$$

and so $s = 150$ m.

Example

A car is being driven at a speed of 20 m/s along a straight road when the driver is aware of a broken down car 50 m in front and blocking the road. If the driver immediately applies the brakes and they give a uniform retardation of 1.5 m/s, can the car stop before hitting the broken down car?

Using equation [11], i.e. $v^2 = u^2 + 2as$, then since the final velocity is 0 the stopping distance required is given by:

$$0 = 20^2 + 2 \times (-1.5) \times s$$

as 133.3 m and so the car does not come to a halt in time.

Example

The velocity v of an object moving along a straight line path is related to the time elapsed t by $v = 10 - 6t$ m/s. Determine (a) the acceleration and (b) the displacement after a time of 4 s.

(a) Acceleration is the rate of change of velocity with time, i.e. dv/dt, thus, differentiating the equation $v = 20 - 6t$ gives an acceleration of $a = dv/dt = -6$ m/s^2.

(b) Since $v = ds/dt$, then integrating gives:

$$\int_0^s ds = \int_0^t v \, dt$$

Thus:

$$s = \int_0^4 (20 - 6t)\, dt = [20t - 3t^2]_0^4 = 32 \text{ m}$$

Revision

1   An object initially at rest is accelerated at a constant 4 m/s² for 8 s. What will be the distance covered in that time?

2   An object has an initial velocity of 3 m/s and 10 s later, following uniform acceleration, has a velocity of 5 m/s. What is the distance covered in that time?

3   A car is accelerated at a uniform 2 m/s² from a velocity of 7.5 m/s until it reaches 22.5 m/s. Calculate the time taken and the distance travelled in that time.

4   The velocity v of an object moving along a straight line path is related to the time elapsed t by $v = 3 - 2t$ m/s. Determine (a) the acceleration and (b) the displacement after a time of 1 s.

4.2.2 Vertical motion under gravity

When an object freely falls under gravity it moves, when the effects of the medium in which it is falling are neglected, with a uniform acceleration called the *acceleration due to gravity*. This varies from one locality to another and at the surface of the earth at sea level is approximately 9.81 m/s². It is customary in considering falling objects to consider upward directions from the point of projection as positive and downward directions as negative. As a consequence, upward directed velocities and accelerations are positive and downward directed velocities and accelerations are negative. Thus, on this convention, the acceleration due to gravity is negative.

Example

An object is projected vertically upwards with an initial velocity of 40 m/s. Calculate (a) the maximum height reached, (b) the time taken to reach the maximum height, (c) the time taken for the object to fall back to its initial point of projection.

(a) Using equation [11], i.e. $v^2 = u^2 + 2as$, with $a = -9.81$ m/s² and the velocity at the greatest height as 0:

$$0 = 40^2 - 2 \times 9.81 \times s$$

Thus the greatest height s is 81.5 m.

(b) Using equation [8], i.e. $v = u + at$, with $a = -9.81$ m/s² and the velocity at the greatest height as 0:

$$0 = 40 - 9.81t$$

Hence the time t is 4.1 s.

(c) The time taken to fall from the greatest height to the initial point of projection is given by equation [9], i.e. $s = ut + \frac{1}{2}at^2$ as:

$$-81.5 = 0 - \frac{1}{2} \times 9.81t^2$$

Hence the time taken to fall is 4.1 s. It is the same time as it took to rise to the maximum height. Thus the total time taken is 8.2 s.

Revision

5   A brick falls off the top of a building. If the height of the building is 20 m, determine the time the brick takes to reach the ground.

6   A ball is thrown vertically upwards with a velocity of 15 m/s. Determine the greatest height reached above the point of projection.

7   A ball is thrown vertically downwards with an initial velocity of 4 m/s from the edge of a cliff. If the ball hits the ground at the base of the cliff after 2 s, determine the height of the cliff.

4.3 Two and three-dimensional linear motion

Displacement, velocity and acceleration are vector quantities and thus can be represented by arrow-headed lines, the length of the line being proportional to the magnitude of the vector and the direction indicated by the arrow being the direction of the vector. As such, they can be added, or subtracted, by the methods used for vectors, e.g. the triangle or parallelogram rules.

Example

An aeroplane sets a course due north with a speed of 200 km/h. However, there is a wind of 50 km/h blowing from the south-west. Determine the actual velocity of the plane.

This example involves the addition of two velocities, 200 km/h in a northerly direction and 50 km/h in a direction from the south-west. Figure 4.13 shows how these velocities can be added by the use of the parallelogram rule. Drawing the parallelogram involves drawing an arrow-headed line to represent one of the velocities, then drawing the arrow-headed line representing the other velocity so that it starts from the start point of the first velocity drawn. The opposite sides of the parallelogram can then be drawn with the resultant velocity being the line drawn as the diagonal from the start points of the two velocities already drawn. The resultant can be determined from a scale drawing or by using the sine and cosine rules. The result is 238 km/h in a direction 8.5° east of north.

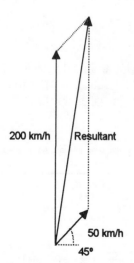

200 km/h   Resultant

50 km/h

45°

Figure 4.13   *Example*

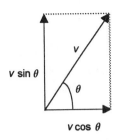

Figure 4.14   *Resolution*

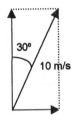

Figure 4.15   *Example*

Revision

8   Determine the resultant velocity when velocities of 16 km/h in an easterly direction and 10 km/h in a direction 38° east of north act on an object.

4.3.1 Resolution

A vector quantity can be resolved, using the parallelogram rule, into two components at right angles to each other. Thus the velocity v in Figure 4.14 can be replaced by the two velocities $v \cos \theta$ and $v \sin \theta$.

Example

Determine the components in an easterly and a northerly direction of a velocity of 10 m/s in a direction 30° east of north.

Figure 4.15 shows the velocity and the components. The component of the velocity in the easterly direction is $10 \cos 60° = 5.0$ m/s and that in the northerly direction is $10 \sin 60° = 8.7$ m/s.

Revision

9   Determine the components in an easterly and a northerly direction of a velocity of 4 m/s in a direction 20° east of north.

4.3.2 Projectiles

Consider an object which is projected with a velocity u at an angle θ above the horizontal (Figure 4.16). The horizontal component of the velocity is $u \cos \theta$ and the vertical component is $u \sin \theta$. There is no gravitational force acting in the horizontal direction and so there is no acceleration imposed on the motion in that direction. The horizontal velocity thus remains, when we neglect air resistance, constant.

> For a projectile: the horizontal motion is with constant velocity but the vertical motion is one of uniform acceleration.

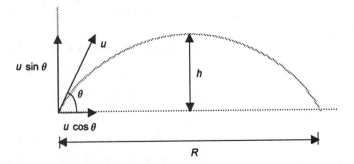

Figure 4.16   *Projectile*

Thus, after a time t from its projection, the horizontal distance x from the start is:

$$x = u \cos \theta \times t \tag{12}$$

There is a vertical force of gravity and so the motion in the vertical direction suffers an acceleration. Thus after a time t, equation [9], i.e. $s = ut + \frac{1}{2}at^2$, gives for the vertical distance y travelled:

$$y = u \sin \theta \times t - \frac{1}{2}gt^2 \tag{13}$$

where g is the acceleration due to gravity.

The maximum height h reached by the projectile is when the vertical component of the velocity is zero. Thus, using equation [11], i.e. $v^2 = u^2 + 2as$, gives:

$$0 = (u \sin \theta)^2 - 2gh$$

Hence:

$$\boxed{\text{greatest height reached } h = \frac{u^2 \sin^2 \theta}{2g}} \tag{14}$$

At the end of the flight the vertical displacement is zero. Thus, if T is the time of flight, equation [9], i.e. $s = ut + \frac{1}{2}at^2$, gives:

$$0 = u \sin \theta \times T - \frac{1}{2}gT^2$$

Hence:

$$\boxed{\text{time of flight } T = \frac{2u \sin \theta}{g}} \tag{15}$$

The horizontal distance, i.e. the range, R covered in time T is the horizontal velocity multiplied by the time and thus:

$$R = u \cos \theta \times \frac{2u \sin \theta}{g} = \frac{2u^2 \sin \theta \cos \theta}{g}$$

Since $2 \cos \theta \sin \theta = \sin 2\theta$, then:

$$\boxed{\text{range } R = \frac{u^2 \sin 2\theta}{g}} \tag{16}$$

For maximum range, $\sin 2\theta = 1$, i.e. $2\theta = 90°$ and $\theta = 45°$, and thus the maximum range is u^2/g. For a given speed of projection there are, in general, two possible angles of projection to give a particular range. Equation [16] can be written as:

$$\sin 2\theta = \frac{Rg}{u^2} \tag{17}$$

For a given value of sin 2θ there are two values of the angle less than 180°. If 2θ is one value then the other value is $180° - 2\theta$.

Example

A gun fires a shell with a velocity of 400 m/s at an elevation above the horizontal of 35°. Determine the maximum height reached by the shell and its range.

Using equation [14]:

$$h = \frac{u^2 \sin^2\theta}{2g} = \frac{400^2 \sin^2 35°}{2 \times 9.81} = 2683 \text{ m}$$

Using equation [16]:

$$R = \frac{u^2 \sin 2\theta}{g} = \frac{400^2 \sin 70°}{9.81} = 15\,326 \text{ m}$$

Example

A stone is thrown from the edge of a cliff with a velocity of 50 m/s at an elevation above the horizontal of 15° (Figure 4.17). If the stone strikes the sea at a point 240 m from the foot of the cliff, determine the height of the cliff.

The horizontal velocity is 50 cos 15° m/s and thus the time taken to travel 240 m is:

$$\text{time} = \frac{240}{50 \cos 15°} = 4.97 \text{ s}$$

The vertical motion has an initial velocity of 50 sin 15° m/s and thus, using equation [9], i.e. $s = ut + \frac{1}{2}at^2$:

$$h = 50 \sin 15° \times 4.97 - \frac{1}{2} \times 9.81 \times 4.97^2 = -56.8 \text{ m}$$

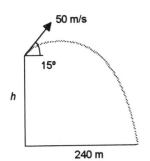

50 m/s

15°

h

240 m

Figure 4.17   *Example*

Revision

10 A projectile is given an initial velocity of 10 m/s at an elevation from the horizontal of 30°. Determine the greatest height reached and the range.

11 What are the angles of projection at which a projectile with an initial velocity magnitude of 10 m/s can have a range of 9 m?

12 A ball is thrown with an initial velocity of 25 m/s at an elevation to the horizontal of 40°. Determine the greatest height reached, the time of flight and the range.

4.4 Angular motion

The following are basic terms used to describe angular motion.

1   *Angular displacement*
The angular displacement θ is the angle swept out by the rotation (Figure 4.18) and is measured in radians. One complete rotation through 360° is an angular displacement of 2π rad.

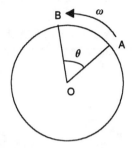

Figure 4.18   *Angular motion*

2   *Angular velocity*
The average angular velocity over some time interval is the change in angular displacement during that time divided by the time:

$$\text{average angular velocity} = \frac{\text{change in angular displacement}}{\text{time for change}} \qquad [18]$$

The unit is rad/s. Thus if a body is rotating at f revolutions per second then it completes $2\pi f$ rad in 1 s and so has an average angular velocity given by:

$$\omega = 2\pi f \qquad [19]$$

The instantaneous angular velocity ω is the change in angular displacement with time when the time interval tends to zero. It can be expressed as:

$$\omega = \frac{d\theta}{dt} \qquad [20]$$

3   *Angular acceleration*
The average angular acceleration over some time interval is the change in angular velocity during that time divided by the time:

$$\text{average angular acceleration} = \frac{\text{change in angular velocity}}{\text{time for change}} \qquad [21]$$

The unit is rad/s². The instantaneous angular acceleration a is the change in angular velocity with time when the time interval tends to zero. It can be expressed as:

$$a = \frac{d\omega}{dt} \qquad [22]$$

4.4.1 Motion with constant angular acceleration

For a body rotating with a constant angular acceleration a, when the angular velocity changes uniformly from ω_0 to ω in time t, as in Figure 4.19, equation [21] gives:

$$a = \frac{\omega - \omega_0}{t}$$

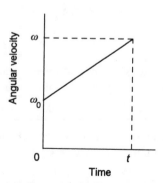

Figure 4.19   *Uniformly accelerated motion*

and hence:

$$\omega = \omega_0 + at \qquad\qquad [23]$$

The average angular velocity during this time is $\frac{1}{2}(\omega + \omega_0)$ and thus if the angular displacement during the time is θ:

$$\frac{\theta}{t} = \frac{\omega + \omega_0}{2}$$

Substituting for ω using equation [23]:

$$\frac{\theta}{t} = \frac{\omega_0 + at + \omega_0}{2}$$

Hence:

$$\theta = \omega_0 t + \frac{1}{2}at^2 \qquad\qquad [24]$$

Squaring equation [23] gives:

$$\omega^2 = (\omega_0 + at)^2 = \omega_0^2 + 2a\omega_0 + a^2t^2 = \omega_0^2 + 2a(\omega_0 + \tfrac{1}{2}at^2)$$

Hence, using equation [24]:

$$\omega^2 = \omega_0^2 + 2a\theta \qquad\qquad [25]$$

Example

An object which was rotating with an angular velocity of 4 rad/s is uniformly accelerated at 2 rad/s. What will be the angular velocity after 3 s?

Using equation [23]:

$$\omega = \omega_0 + at = 4 + 2 \times 3 = 10 \text{ rad/s}$$

Example

The blades of a fan are uniformly accelerated and increase in frequency of rotation from 500 to 700 rev/s in 3.0 s. What is the angular acceleration?

Since $\omega = 2\pi f$, equation [23] gives:

$$2\pi \times 700 = 2\pi \times 500 + a \times 3.0$$

Hence $a = 419 \text{ rad/s}^2$.

Example

A flywheel, starting from rest, is uniformly accelerated from rest and rotates through 5 revolutions in 8 s. What is the angular acceleration?

The angular displacement in 8 s is $2\pi \times 5$ rad. Hence, using equation [24], i.e. $\theta = \omega_0 t + \frac{1}{2}at^2$:

$$2\pi \times 5 = 0 + \frac{1}{2}a \times 8^2$$

Hence the angular acceleration is 0.98 rad/s².

Revision

13   A flywheel rotating at 3.5 rev/s is accelerated uniformly for 4 s until it is rotating at 9 rev/s. Determine the angular acceleration and the number of revolutions made by the flywheel in the 4 s.

14   A flywheel rotating at 20 rev/min is accelerated uniformly for 10 s until it is rotating at 40 rev/min. Determine the angular acceleration and the number of revolutions made by the flywheel in the 10 s.

4.4.2 Relationship between linear and angular motion

Consider the rotating line OA in Figure 4.20. When OA rotates through angle θ to OB, point A moves in a circular path and the distance moved by the point A round the circumference is:

$$s = r\theta \qquad [26]$$

If the point is moving with constant angular velocity ω then in time t the angle rotated will be ωt. Thus:

$$s = r\omega t$$

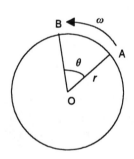

Figure 4.20   *Angular motion*

But s/t is the linear speed v of point A round the circumference. Hence:

$$v = r\omega \qquad [27]$$

We could have derived equation [27] without considering constant angular velocity and dealing with the instantaneous angular velocity. Thus if we differentiate equation [26] with respect to time:

$$\frac{ds}{dt} = r\frac{d\theta}{dt}$$

ds/dt is the linear velocity and $d\theta/dt$ the angular velocity. Thus equation [27] is obtained.

If we differentiate equation [27] with respect to time:

$$\frac{dv}{dt} = r\frac{d\omega}{dt}$$

But dv/dt is the linear acceleration a and $d\omega/dt$ is the angular acceleration α. Thus:

$$a = r\alpha \qquad\qquad\qquad\qquad [28]$$

Example
What is the peripheral velocity of a point on the rim of a wheel when it is rotating at 3 rev/s and has a radius of 200 mm?

Using equation [27]:

$$v = r\omega = 0.200 \times 2\pi \times 3 = 3.8 \text{ m/s}$$

Example
The wheels of a car have a diameter of 700 mm. If they increase their rate of rotation from 50 rev/min to 1100 rev/min in 40 s, what is the angular acceleration of the wheels and the linear acceleration of a point on the tyre tread?

Using equation [23], i.e. $\omega = \omega_0 + \alpha t$, then:

$$2\pi \times \frac{1100}{60} = 2\pi \times \frac{50}{60} + a \times 40$$

Hence the angular acceleration is 2.75 rad/s². Using equation [28], i.e. $a = r\alpha$, then:

$$a = 0.350 \times 2.75 = 0.96 \text{ m/s}$$

Revision

15 The linear speed of a belt passing round a pulley wheel of radius 150 mm is 20 m/s. If there is no slippage of the belt on the wheel, how many revolutions per second are made by the wheel?

16 A car has wheels of diameter 550 mm and is travelling along a straight road with a constant speed of 20 m/s. What is the angular velocity of the wheel?

17 A cord is wrapped around a wheel of diameter 400 mm which is initially at rest (Figure 4.21). When the cord is pulled, a tangential acceleration of 4 m/s² is applied to the wheel. What is the angular acceleration of the wheel?

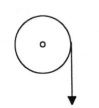

Figure 4.21   *Problem 17*

4.4.3 Gears and belts

Two intermeshed gears can be considered as basically two disks in contact   (Figure 4.22) and for which no slippage occurs when one of them is rotated and causes the other to rotate, the gear teeth being to prevent slippage. At the point of contact between the two gears we must have the same tangential velocity. Thus, using equation [27], i.e. $v = r\omega$, we must have:

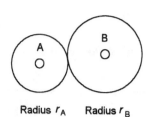

Radius r_A Radius r_B

Figure 4.22   *Two gears*

$$r_A\omega_A = r_B\omega_B \qquad\qquad [29]$$

Likewise, at the point of contact the tangential acceleration will be the same. Thus, using equation [28], i.e. $a = r\alpha$, we must have:

$$r_A a_A = r_B a_B \qquad\qquad [30]$$

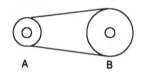

A   B

Figure 4.23   *Belt drive*

Belt drives (Figure 4.23) are similar to gears except that the motion is transferred between the two pulleys by a belt rather than the direct contact that occurs with gears. Equations [29] and [30] thus apply.

Example

Gear A is in mesh with gear B, gear A having a radius of 20 mm and gear B a radius of 50 mm. If A starts from rest and has a constant angular acceleration of 2 rad/s², determine the time taken for B to reach an angular velocity of 50 rad/s.

Using equation [30], i.e. $r_A \dot{a}_A = r_B a_B$:

$$0.020 \times 2 = 0.050 \times a_B$$

Thus $a_B = 0.8$ rad/s². Using equation [23], i.e. $\omega = \omega_0 + at$:

$$50 = 0 + 0.8t$$

Hence $t = 62.5$ s.

Example

Pulley A drives pulley B by means of a belt drive, A having a radius of 100 mm and B a radius of 300 mm. Pulley A starts from rest and accelerates uniformly at 15 rad/s² for 12 s. Determine the number of revolutions of pulley B in that 12 s time interval.

Using equation [30], i.e. $r_A a_A = r_B a_B$:

$$0.100 \times 12 = 0.300 \times a_B$$

Thus $a_B = 4$ rad/s^2. Using equation [24], i.e. $\theta = \omega_0 t + \frac{1}{2}at^2$, then:

$$\theta = 0 + \frac{1}{2} \times 4 \times 12^2 = 72 \text{ rad}$$

The number of revolutions is thus $72/2\pi = 11.5$.

Revision

18   Gear A and gear B are intermeshed, A having a radius of 200 mm and B a radius of 150 mm. Gear A starts from rest and rotates with a constant angular acceleration of 2 rad/s^2. Determine the angular velocity and angular acceleration of B after A has completed 10 rev.

19   A 200 mm diameter drive pulley rotates at 5 rev/s. It drives, via a belt drive, another pulley. What diameter will this need to be if it is required to rotate a shaft at 2.5 rev/s?

4.4.4 Combined linear and angular motion

Consider objects which have both a linear motion and angular motion, e.g. a rolling wheel. For a wheel of radius r which is rolling, without slip, along a straight path (Figure 4.24), when the wheel rotates and rolls its centre moves from C to C′ then a point on its rim moves from O to O′. The distance CC′ equals OO′. But OO′ = $r\theta$. Thus:

horizontal distance moved by the wheel $x = r\theta$ [31]

If this movement occurs in a time t, then differentiating equation [31]:

$$\frac{dx}{dt} = r\frac{d\theta}{dt}$$

and thus:

horizontal velocity $v_x = r\omega$ [32]

where ω is the angular velocity of the wheel.

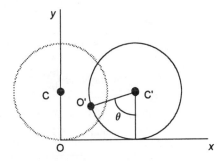

Figure 4.24   *Rolling wheel*

Differentiating equation [32] gives:

$$\frac{dv_x}{dt} = r\frac{d\omega}{dt}$$

and thus:

horizontal acceleration $= ra$ [33]

where a is the angular acceleration.

4.5 Force and linear motion

In considering the effects of force on the motion of a body we use Newton's laws.

> *Newton's laws* can be expressed as:
>
> *First law*
> A body continues in its state of rest or uniform motion in a straight line unless acted on by a force.
>
> *Second law*
> The rate of change of momentum of a body is proportional to the applied force and takes place in the direction of the force.
>
> *Third law*
> When a body A exerts a force on a body B, B exerts an equal and opposite force on A (this is often expressed as: to every action there is an opposite and equal reaction).

Thus the first law indicates that if we have an object moving with a constant velocity there can be no resultant force acting on it. If there is a resultant force, then the second law indicates that there will be a change in momentum with:

force $F \propto$ rate of change of momentum, i.e. $\dfrac{d(mv)}{dt}$ [34]

For a mass m which does not change with time:

force $F \propto m\dfrac{dv}{dt}$

dv/dt is the acceleration. The units are chosen so that the constant of proportionality is 1 and thus the second law can be expressed as:

$$F = ma$$ [35]

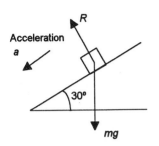

Figure 4.25   *Example*

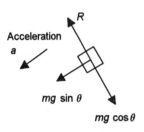

Figure 4.26   *Example*

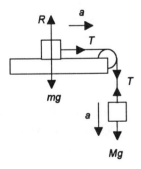

Figure 4.27   *Example*

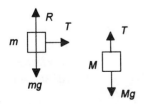

Figure 4.28   *Example*

An object resting in equilibrium on a table has zero acceleration and thus, according to Newton's second law, the net force on the object must be zero. Hence the weight of the body which acts downwards on the table must be exactly balanced by an upward acting force exerted by the table on the body, this force being termed the reaction.

Force is a vector quantity and thus, when adding or subtracting forces, the parallelogram or triangle laws have to be used. A common method in problems involving more than one force is to resolve the forces acting on an object into the horizontal and vertical directions. Then the forces acting in the same directions can be added.

Example

A body of mass 3 kg slides down an smooth plane inclined at 30° to the horizontal. Determine the acceleration of the body and the normal reaction exerted by the plane on the body.

Figure 4.25 shows the forces acting on the body; because the plane is termed as being smooth we assume there are no frictional forces. The weight of the body is mg, where g is the acceleration due to gravity. R is the reaction of the plane arising from the body pressing down on it. It is at right angles to the plane. If we resolve the weight of the body into a force component parallel to the plane and one at right angles to it, then the free body diagram for the forces acting on the body is as shown in Figure 4.26. Thus the force accelerating the body down the plane is $mg \sin \theta$, where θ is the angle of the plane from the horizontal. Thus applying equation [35], i.e. $F = ma$:

$$ma = mg \sin \theta$$

and so the acceleration a = g sin θ = 9.81 × sin 30° = 4.9 m/s². The reaction R is equal to $mg \cos \theta$ since there is no motion in the direction at right angles to the plane. Thus R = 3 × 9.81 × cos 30° = 25.5 N.

Example

A block of mass 6 kg rests on a horizontal table top and is connected by a light inextensible string that passes over a pulley at the edge of the table to another block of mass 5kg which is hanging freely (Figure 4.27). Determine the acceleration of the system and the tension in the string.

Figure 4.28 shows the free body diagrams for the two masses. For the block on the table, the force causing acceleration is T and thus:

$$T = ma = 6a \qquad [36]$$

For the hanging block, the net force causing acceleration is $Mg - T$ and thus:

$$5g - T = 5a \qquad [37]$$

Using equation [36] to eliminate T in equation [37] gives:

$$5g - 6a = 5a$$

and so $a = 5g/11 = 4.46$ m/s². Equation [36] thus gives the tension as $T = 6a = 6 \times 4.46 = 26.76$ N.

Example

Figure 4.29 shows a pulley system. A is a fixed smooth pulley and pulley B, also smooth, has a mass of 4 kg. A load of mass 5 kg is attached to the free end of the rope. Determine the acceleration of the load and the tension in the rope when the load is released.

When the load is released, it will move in a particular time through twice the distance the pulley B will move through. Thus if a is the acceleration of pulley B, then the acceleration of the load will be $2a$. For the load, the net downward force is $5g - T$ and thus:

$$5g - T = 5 \times 2a$$

For pulley B, the net upward force is $2T - 4g$ and thus:

$$2T - 4g = 4a$$

Hence, eliminating T from these equations gives $a = 2.45$ m/s². Pulley B has thus an acceleration of 2.45 m/s² upwards and the load an acceleration of 4.90 m/s² downwards. The tension on the string is 24.6 N.

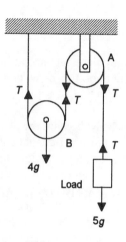

Figure 4.29   *Example*

Revision

20 A train engine exerts a force of 35 kN on a train of mass 240 Mg and pulls it up a slope of 1 in 120 against a resistive force of 14 kN. Determine the acceleration up the slope of the train.

21 Masses of 3 kg and 2 kg are connected by a light inextensible string which passes over a smooth, fixed pulley. Determine the acceleration of the masses and the tension in the string.

22 An object of mass 40 kg is on a smooth incline at an elevation of 20° to the horizontal. It is connected by a cord over a pulley at the bottom of the slope to a vertically hanging load of 30 kg and slides down the slope against a resistance of 70 N. Determine its acceleration. The pulley can be assumed to be smooth and the cord as having negligible mass and being inextensible.

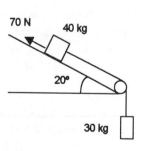

Figure 4.30   *Revision problem 22*

4.5.1 Friction

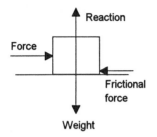

Figure 4.31 *Frictional force*

When there is a tendency for one surface to slide over another, a frictional force occurs which acts along the common tangent at the point of contact and so at right angles to the contact reaction force. The frictional force is always acting in such a direction as to oppose the motion of sliding. Thus when an object, resting on a horizontal surface is acted on by a force *F*, there will be an opposing frictional force (Figure 4.31). If the force is gradually increased from zero, the object does not move until the applied force reaches some particular value. When the object is not moving, though a force is being applied, we must have no net force and so the frictional force must be opposite and equal to the applied force and thus increase as the applied force increases. When the object starts to slide under the action of the applied force then there is a net force and so the applied force is greater than the frictional force, the frictional force must have stopped increasing and reached a maximum or limiting value.

The area number of basic laws governing friction:

1   The frictional force is independent of the areas in contact.

2   The frictional force depends on the nature of the materials in contact.

3   The limiting frictional force *F* is proportional to the normal reaction force *N* between the two surfaces in contact, the constant of proportionality being termed the *coefficient of friction μ*.

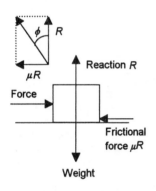

Figure 4.32 *Motion along a horizontal plane*

$$F = \mu R \qquad [38]$$

Consider motion along a horizontal plane (Figure 4.32). The frictional force is μR and thus the resultant of the frictional force and the normal reaction force is a force at an angle ϕ to the reaction force, this angle being termed the *angle of friction* since $\tan \phi = \mu R/R$ and so :

$$\tan \phi = \mu \qquad [39]$$

Example

An object of mass *m* rests on a plane at an elevation of θ to the horizontal (Figure 4.33). Determine the force *P* parallel to the plane that is needed to push the object up the plane with an acceleration *a* if the coefficient of friction is μ.

The forces acting parallel to the plane are the applied force *P*, the frictional force *F* and the component of the weight acting down the plane, i.e. *mg* sin θ. Thus:

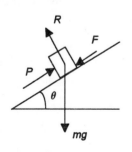

Figure 4.33   *Example*

net force down the plane = $P - F - mg \sin \theta$

Since $F = \mu R$ and $R = mg \cos \theta$, then:

$$P - \mu mg \cos \theta - mg \sin \theta = ma$$

and so:

$$a = (P/m) - \mu g \cos \theta - g \sin \theta$$

Example

An object of mass m rests on a plane at an elevation of θ to the horizontal (Figure 4.34). Determine the angle to which the plane can be tilted until the block just begins to slide down it.

The force parallel to the plane is the component of the weight in that direction, i.e. $mg \sin \theta$, and the frictional force μR. The block will just begin to slide down the plane when:

$$mg \sin \theta = \mu R$$

But $R = mg \cos \theta$, thus the condition for motion to start is:

$$mg \sin \theta = \mu mg \cos \theta$$

and so it is:

$$\tan \theta = \mu$$

Example

An object of mass 0.5 kg rests on a rough plane inclined at 40° to the horizontal. If the coefficient of friction is 0.5, determine the smallest force that can be applied parallel to the plane to prevent it from sliding down the plane.

The net force parallel to the plane must be zero. The frictional force acts up the plane since the object is on the point of sliding down the plane. We have $F = \mu R = \mu mg \cos \theta$. Thus:

$$P + F = mg \sin \theta$$

$$P + \mu mg \cos \theta = mg \sin \theta$$

and so:

$$P = mg(\sin \theta - \mu \cos \theta)$$

$$= 0.5 \times 9.81(\sin 40° - 0.5 \cos 40°) = 1.27 \text{ N}$$

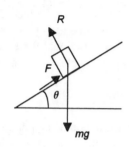

Figure 4.34   *Example*

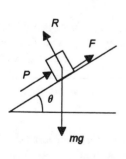

Figure 4.35   *Example*

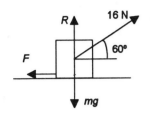

Figure 4.36   *Example*

Example

An object of mass 2.5 kg is on a horizontal plane and being pulled by a string inclined at 60° to the horizontal (Figure 4.36). If the object is just on the point of sliding along the plane when the tension in the string is 16 N, what is the coefficient of friction?

For the vertical forces to be in equilibrium we must have:

$$R + 16 \sin 60° = mg$$

Thus $R = 2.5 \times 9.81 - 16 \sin 60° = 10.7$ N. When the object is just on the point of moving the net horizontal force must be zero. Thus:

$$16 \cos 60° = F = \mu R = \mu \times 10.7$$

Hence the coefficient of friction is 0.75.

Revision

23 An object of mass 6 kg rests in equilibrium on a rough plane inclined at 30° to the horizontal. Determine the coefficient friction between the object and the plane.

24 An object of mass 0.5 kg is on a rough plane at an elevation of 20° to the horizontal and is acted on by a force of 6 N up and parallel to the plane. If the coefficient of friction is 0.7, what is the acceleration of the object up the plane?

25 An object of weight 24 N rests on a rough plane inclined at 30° to the horizontal and is held by the tension in a string parallel to the plane. Determine the tension in the string if the coefficient of friction is 0.2.

26 An object of mass 60 kg is at rest on a rough plane which is inclined at 45° to the horizontal. Determine the force needed to maintain the object at rest if the force is applied (a) parallel to the plane, (b) horizontally. The coefficient of friction is 0.25.

4.6 Torque and angular motion

Consider a force F acting on a small element of mass δm of a rigid body, the element being a distance r from the axis of rotation at O (Figure 4.37). The torque T acting on the element is Fr and since $F = \delta m \times a$ we can express the torque as:

$$\text{torque } T = \delta m \times ar$$

But $a = r\alpha$, where α is the angular acceleration. Thus $T = \delta m \times r^2\alpha$. We can express this as:

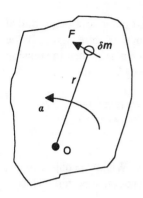

Figure 4.37   *Rotation of a rigid body*

torque $T = Ia$ [40]

where I is termed the *moment of inertia* and for the small element of mass m at radius r is given by:

moment of inertia $I = \delta m \times r^2$ [41]

Equation [40] gives the torque required to give an angular acceleration to just an element of the mass of the rigid body. The torque required to give a rotational acceleration for the entire rigid body will thus be the sum of the torques required to accelerate each element of mass in the body and given by equation [40] when we consider the sum of the moments of inertia of the entire body. Thus:

torque to give body angular acceleration $T = Ia$ [42]

where:

moment of inertia of body $I = \int r^2 \, dm$ [43]

4.6.1 Moment of inertia

For a small mass at the end of a light pivoted arm with radius of rotation r, we can consider the entire mass of the body to be located at the same distance r and so the moment of inertia is given by equation [43] as:

$$I = mr^2$$ [44]

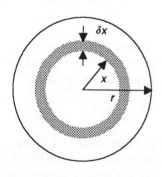

Figure 4.38   *Uniform disc*

For a uniform disc of radius r and total mass m, the moment of inertia about an axis through the disc centre can be found by considering the disc to be composed of a number of rings (Figure 4.38). For a ring of thickness δx and radius x, its area is effectively its circumference multiplied by the thickness and so is $2\pi x \, \delta x$. Since the mass per unit area of the disc is $m/\pi r^2$ then the mass δm of this ring element is:

$$\delta m = \frac{m}{\pi r^2} 2\pi x \delta x = \frac{2m}{r^2} x \delta x$$

Thus equation [43] gives for the moment of inertia of the disc:

$$I = \int_0^r x^2 \, dm = \int_0^r \frac{2m}{r^2} x^3 \, dx$$

$$I = \tfrac{1}{2} mr^2$$ [45]

In a similar manner we can derive the moments of inertia of other bodies. Figure 4.39 gives the moments of inertia for some commonly encountered bodies.

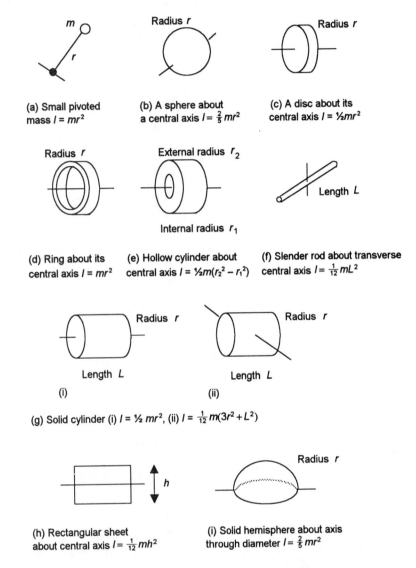

(a) Small pivoted mass $I = mr^2$

(b) A sphere about a central axis $I = \frac{2}{5}mr^2$

(c) A disc about its central axis $I = \frac{1}{2}mr^2$

(d) Ring about its central axis $I = mr^2$

(e) Hollow cylinder about central axis $I = \frac{1}{2}m(r_2^2 - r_1^2)$

(f) Slender rod about transverse central axis $I = \frac{1}{12}mL^2$

(g) Solid cylinder (i) $I = \frac{1}{2}mr^2$, (ii) $I = \frac{1}{12}m(3r^2 + L^2)$

(h) Rectangular sheet about central axis $I = \frac{1}{12}mh^2$

(i) Solid hemisphere about axis through diameter $I = \frac{2}{5}mr^2$

Figure 4.39   *Moments of inertia about the indicated axes*

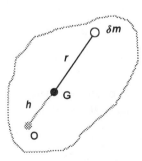

Figure 4.40   *Moment of inertia about a parallel axis*

The value of the moment of inertia for a particular body depends on the axis about which it is calculated. If I_G is the moment of inertia about an axis through the centre of mass G, consider what the moment of inertia would be about a parallel axis through O, a distance h away (Figure 4.40). The moment of inertia of an element dm of the mass about the axis through O is:

$$\text{moment of inertia of element} = (r + h)^2 \, \delta m$$

$$= (r^2 + h^2 + 2rh) \, \delta m$$

The total moment of inertia about the axis through O is thus:

$$I_O = \int r^2 \, dm + \int h^2 \, dm + \int 2rh \, dm$$

The moment of inertia about an axis through the centre of mass G is $I_G = \int r^2 \, dm$. The integral $\int h^2 \, dm = mh^2$. The integral $2h\int r \, dm$ is the total moment of the mass about the axis through the centre of mass and is thus zero. Hence:

$$I_O = I_G + mh^2 \qquad\qquad\qquad [46]$$

This is known as the *theorem of parallel axes*.

It is sometimes necessary to obtain the moment of inertia of a composite body about some axis. This can be done by summing the moments of inertia of each of the elemental parts about the axis, using the parallel axis theorem where necessary. This is illustrated in one of the following examples.

Consider replacing a body by one of the same mass but it all concentrated at just one point, the point being so located that the same moment of inertia is obtained. If the total mass of a body is m and we consider it all to be concentrated at a small point a distance k from the pivot axis then its moment of inertia would be

$$I = mk^2 \qquad\qquad\qquad [47]$$

k is called the *radius of gyration*. Thus, for a disc of radius r and mass m the moment of inertia about its central axis is $\frac{1}{2}mr^2$ and thus the radius of gyration for the disc is $k^2 = \frac{1}{2}r^2$ and so $k = r/\sqrt{2}$.

Example

Determine the moment of inertia about an axis through its centre of a disc of radius 100 mm and mass 200 g.

The moment of inertia of the disc is given by:

$$I = \tfrac{1}{2}mr^2 = \tfrac{1}{2} \times 0.200 \times 0.100^2 = 1.0 \times 10^{-3} \text{ kg m}^2$$

Example

Determine the moment of inertia about an axis of an object having a mass of 2.0 kg and a radius of gyration from that axis of 300 mm.

The moment of inertia is given by equation [47] as:

$$I = mk^2 = 2.0 \times 0.300^2 = 0.18 \text{ kg m}^2$$

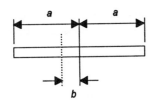

Figure 4.41   *Example*

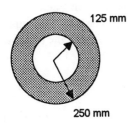

Figure 4.42   *Example*

Example

Determine the moment of inertia of a slender rod of mass m about an axis at right angles to its length of $2a$ and a distance b from its centre (Figure 4.41).

The moment of inertia of a slender rod of length L about an axis through its centre is $mL^2/12$. Thus, for $L = 2a$, we have

$$I = \frac{ma^2}{3}$$

Using the parallel axis theorem, the moment of inertia about a parallel axis a distance b from the centre is thus:

$$I = \frac{ma^2}{3} + mb^2$$

Example

Determine the moment of inertia of the plate shown in Figure 4.42 about an axis at right angles to it and through its centre. The plate is 10 mm thick and has a radius of 250 mm and contains a central hole with a radius of 125 mm. The material has a density of 8000 kg/m³.

This can be tackled by considering the plate as a composite body, i.e. a solid disc of radius 250 mm minus a disc of radius 125 mm. For the 250 mm diameter disc, the moment of inertia is:

$$I = \tfrac{1}{2}MR^2 = \tfrac{1}{2} \times \pi \times 0.250^2 \times 0.010 \times 8000 \times 0.250^2$$

$$= 0.491 \text{ kg m}^2$$

For the 125 mm diameter disc:

$$I = \tfrac{1}{2}MR^2 = \tfrac{1}{2} \times \pi \times 0.125^2 \times 0.010 \times 8000 \times 0.125^2$$

$$= 0.031 \text{ kg m}^2$$

Thus the moment of inertia of the disc with the hole is $0.491 - 0.031 = 0.460$ kg m².

Revision

27 Determine the moment of inertia of an object having a mass of 4 kg and a radius of gyration of 200 mm.

28 Determine the moment of inertia of a sphere about an axis through its centre if it has a radius of 200 mm and is made of material with a density of 8000 kg/m³.

29   A disc with an outside radius of 250 mm and a mass of 11.78 kg has a moment of inertia of 0.460 kg m² about an axis at right angles to its plane and through its centre. Determine its moment of inertia about an axis at right angles to its plane and passing through a point on its circumference.

30   Determine the moment of inertia of a disc, about an axis perpendicular to the face of the disc and through its centre, which has a constant thickness of 50 mm, a diameter of 750 mm and a central hole of diameter 150 mm. The disc material has a density of 7600 kg/m³.

4.6.2 Rotational problems

The following examples illustrate how rotational problems involving moments of inertia can be tackled.

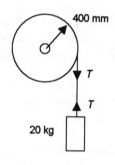

Figure 4.43   *Example*

Example

A drum with a mass of 60 kg, radius of 400 mm and radius of gyration of 250 mm has a rope of negligible mass wrapped round it and attached to a load of 20 kg (Figure 4.43). Determine the angular acceleration of the drum when the load is released.

For the forces on the load we have:

$$20g - T = 20a \tag{48}$$

For the drum, the torque is $T \times 0.400$ and thus, using torque $= Ia$:

$$T \times 0.400 = Ia$$

Since $I = mk^2 = 60 \times 0.250^2$ then:

$$T \times 0.400 = 60 \times 0.250^2 \times a \tag{49}$$

The point of contact between the rope and the drum will have the same tangential acceleration and thus, using $a = ra$:

$$a = 0.400a$$

Thus equation [48] can be written as:

$$20g - T = 20 \times 0.400a$$

and so, eliminating T between this equation and equation [49]:

$$(20g - 20 \times 0.400a) \times 0.400 = 60 \times 0.250^2 \times a$$

The angular acceleration is thus 11.3 rad/s².

Example

A record turntable is a uniform flat plate of radius 120 mm and mass 0.25 kg. What torque is required to uniformly accelerate the turntable to 33.3 rev/min in 2 s?

The angular acceleration required is given by $\omega = \omega_0 + at$ as:

$$2\pi \times \frac{33.3}{60} = 0 + a \times 2$$

Hence $a = 1.74$ rad/s^2. The moment of inertia of the turntable is:

$$I = \tfrac{1}{2}mr^2 = \tfrac{1}{2} \times 0.25 \times 0.120^2 = 1.8 \times 10^{-3} \text{ kg m}^2$$

The torque required to accelerate the turntable is thus:

$$T = Ia = 1.8 \times 10^{-3} \times 1.74 = 3.13 \times 10^{-3} \text{ N m}$$

Example

A solid sphere rolls, without slipping, down a rough inclined plane which is at an elevation of θ to the horizontal (Figure 4.44). Determine its acceleration.

For the linear motion of the centre of mass of the sphere, the forces acting parallel to and down the plane are:

$$mg \sin \theta - F = ma \qquad\qquad [50]$$

where a is the linear acceleration down the plane and F is the frictional force. As the sphere descends it rotates about its centre. The torque giving this rotation is that due to the frictional force and is thus Fr, where r is the radius of the sphere. Thus:

$$Fr = Ia$$

where a is the angular acceleration of the sphere. The moment of inertia I of the sphere is $2mr^2/5$. Thus:

$$Fr = \tfrac{2}{5}mr^2 \times a$$

But $a = ra$ and so:

$$a = ra = r\frac{5F}{2mr} = \frac{5F}{2m}$$

Using equation [50] and substituting for F gives:

$$mg \sin \theta - \frac{2ma}{5} = ma$$

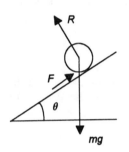

Figure 4.44   *Example*

Hence:

$$a = \tfrac{5}{7}g\sin\theta$$

Example

A cable drum rests on a rough horizontal surface (Figure 4.45). It has a moment of inertia I about its axis of rotation and a mass m. The outer part of the drum has a radius r_2 and the radius round which the cable is wound is r_1. Determine the linear acceleration of the drum when the cable is pulled with a horizontal force of P and the drum rolls without slipping.

For the horizontal linear motion of the centre of mass of the cable drum:

$$P - F = ma \tag{51}$$

where F is the frictional force and a the linear acceleration. For vertical equilibrium we have:

$$R = mg$$

The drum rotates under the net action of the torque applied by the cable and that resulting from friction between the drum and the horizontal surface. Thus:

$$Pr_1 - Fr_2 = I\alpha$$

where α is the angular acceleration of the drum. Using equation [51] to eliminate F gives:

$$Pr_1 - (P - ma)r_2 = I\alpha$$

$$P(r_2 - r_1) = mar_2 - I\alpha$$

Since with no slip we have $a = -r_2\alpha$, the minus sign being because the direction of the acceleration for the drum is in the opposite direction to that for the cable, then:

$$P(r_2 - r_1) = mar_2 + Ia/r_2$$

$$a = \frac{P(r_2 - r_1)}{mr_2 + I/r_2}$$

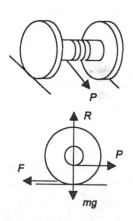

Figure 4.45   *Example*

Revision

31   A disc has a mass of 2.5 kg and a radius of gyration of 100 mm. Determine the torque required to give the disc an angular acceleration of 15 rad/s².

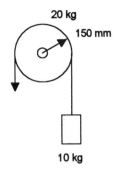

20 kg

150 mm

10 kg

Figure 4.46   *Revision problem 32*

32 A uniform pulley has a mass of 20 kg and a radius of 150 mm. A rope passes over it and is attached to a load of 10 kg (Figure 4.46). What force must be applied to the other end of the rope to give the load an upward acceleration of 0.2 m/s?

33 A uniform cylinder rolls, without slipping, from rest down a rough inclined plane which is at an elevation of 30° to the horizontal. Determine its linear acceleration down the plane.

34 A uniform cylinder of mass 6 kg and radius 100 mm is made to roll without slipping along a rough horizontal surface as a consequence of a horizontal force of 80 N being applied along its central axis. Determine the angular acceleration of the cylinder.

4.7 Linear and angular kinetic energy

When work is done to raise a body of mass m through a vertical height h the work done is the product of the force applied and the distance covered in the direction of the force and so is mgh. This is the energy transferred to the body as a result of its position in the gravitational field and is termed *potential energy*. Thus:

$$\text{potential energy} = mgh \qquad [52]$$

Consider an object which starting from rest is accelerated uniformly over a linear distance s. The work done is the product of the force applied and the distance covered in the direction of the force. But the force is $F = ma$ and so the work done is $Fs = mas$. After the distance s the object has a velocity v which is given by $v^2 = u^2 + 2as$ as $v^2 = 2as$. Thus the work done in accelerating the object to this velocity is:

$$\text{work done} = mas = \tfrac{1}{2}mv^2$$

This is the energy transferred to the body as a result of the work done and is called *kinetic energy*. Thus, for linear motion:

$$\text{kinetic energy} = \tfrac{1}{2}mv^2 \qquad [53]$$

Now consider an object which, starting from rest, is set rotating with an angular acceleration a as a result of a torque T produced by a tangential force F (Figure 4.47). The work done by the force is:

$$\text{work done} = \text{force} \times \text{distance} = F \times \text{arc length} = Fr\theta$$

The torque $T = Fr$ and so the work done is:

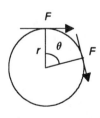

Figure 4.47   *Torque causing rotation*

$$\text{work done} = T\theta \qquad [54]$$

But $T = Ia$ and thus we can write equation [53] as:

work done = $Ia\theta$

As a result of this rotation from rest through angle θ, the object has an angular velocity ω. Using $\omega^2 = \omega_0^2 + 2a\theta$, we have $\omega^2 = 0 + 2a\theta$ and so:

work done = $\frac{1}{2}I\omega^2$

This is the energy transferred to the body as a result of the work done and is called the angular kinetic energy. Thus, for angular motion:

---

angular kinetic energy = $\frac{1}{2}I\omega^2$ [55]

Example

A car of mass 1000 kg is driven up an incline of length 750 m and inclination 1 in 25. Determine the driving force required from the engine if the speed at the foot of the incline is 25 m/s and at the top is 20 m/s and resistive forces can be neglected.

The work done by the driving force F is $F \times 750$ J. There is a change in both potential energy and kinetic energy as a result of the work done. The gain in potential energy is $1000 \times 9.81 \times 750 \sin\theta$, where θ is the angle of elevation of the incline. Since $\sin\theta$ is given as 1/25 then the gain in potential energy is 294.3 kJ. The initial kinetic energy at the foot of the incline is $\frac{1}{2} \times 1000 \times 25^2 = 312.5$ kJ and at the top of the incline is $\frac{1}{2} \times 1000 \times 20^2 = 200$ kJ. Hence there is a loss in kinetic energy of 112.5 kJ in going up the slope. The work done must equal the total change in energy and so:

$F \times 750 = 294.3 \times 10^3 - 112.5 \times 10^3$

and so the driving force is 242.4 N.

Example

A solid cylinder of diameter 25 mm and length 25 mm is allowed to roll down an inclined plane, the plane being at an elevation of 20° to the horizontal. Determine the linear velocity of the cylinder when it has rolled a distance of 1.2 m. The cylinder material has a density of 7000 kg/m³.

The loss in potential energy of the cylinder in rolling down the slope must equal the gain in linear kinetic energy plus the gain in rotational kinetic energy. The loss in potential energy is mgh with m being $\frac{1}{4}\pi \times 0.025^2 \times 0.025 \times 7000 = 0.0859$ kg and $h = 1.2 \sin 20° = 0.410$ m. The loss in potential energy is thus $0.0859 \times 9.81 \times 0.410 = 0.345$ J. The gain in linear kinetic energy is $\frac{1}{2} \times 0.0859v^2$. The gain in rotational kinetic energy is $\frac{1}{2}I\omega^2$, where $I = \frac{1}{2}mr^2 = \frac{1}{2} \times 0.0859 \times 0.0125^2 = 6.71 \times 10^{-6}$ kg m^2 and $\omega = v/r = v/0.0125$. Thus the gain in rotational energy is $\frac{1}{2} \times 6.71 \times 10^{-6} \times (v/0.0125)^2 = 0.0215v^2$. Since the loss in potential energy must equal the total gain in kinetic energy:

$$0.345 = \frac{1}{2} \times 0.0859v^2 + 0.0215v^2$$

and so $v = 2.31$ m/s.

Example

A square trap door, 1.8 m by 1.8 m, has a mass of 10 kg and is at rest in the horizontal position (Figure 4.48). When it is released it rotates about its hinges to strike a rubber door stop when hanging vertically. What will be its angular velocity on impact with the door stop if there is a frictional torque of 20 N m at the hinges?

The centre of mass of the trap door will fall through a vertical height of 0.9 m and so the loss in potential energy will be $10 \times g \times 0.9 = 88.3$ J. The gain in rotational kinetic energy is $\frac{1}{2}I\omega^2$, where the moment of inertia about the centre of mass of the door is $m \times 1.8^2/12$ and, using the theorem of parallel axes, about the edge of the door is $m \times 1.8^2/12 + m \times 0.9^2 = 1.08m$. The work done in rotating the door through 90° against the frictional torque is given by equation [54] as $T\theta = 20 \times \pi/2 = 31.4$ J. The loss in potential energy equals the gain in rotational kinetic energy plus the work done in rotating the door through 90° and so:

$$88.3 = 1.08 \times 10 \times \omega^2 + 31.4$$

and the angular velocity is 2.30 rad/s.

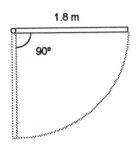

1.8 m

90°

Figure 4.48   *Example*

Example

A load of 50 kg hangs vertically from one end of a rope, the other end being wound round the rim of a drum of diameter 600 mm (Figure 4.49). The drum has a moment of inertia of 30 kg m^2. Initially the system is at rest. A torque of 200 N m is then applied to the drum and the resulting rotation lifts the load. Determine the time taken to lift the load through 30 m.

The movement of the load by 30 m means that the rope has had to be wound up on the drum by that amount. Thus the angle through which the drum must have been rotated is given by the equation arc length = radius × angle rotated as $\theta = 30/0.300 = 100$ rad. The work done by the torque is thus given by equation [54] as $T\theta = 200 \times 100 = 20$ kJ. This work has resulted in the kinetic energy of the drum changing from its initially zero value, the kinetic energy of the load

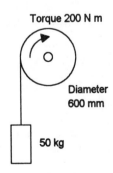

Torque 200 N m

Diameter 600 mm

50 kg

Figure 4.49   *Example*

changing from its initial zero value and the potential energy of the mass changing. Thus:

$$20\,000 = \tfrac{1}{2}I\omega^2 + \tfrac{1}{2}mv^2 + mgh$$

$$= \tfrac{1}{2} \times 30 \times \omega^2 + \tfrac{1}{2} \times 50 \times v^2 + 50 \times 9.81 \times 30$$

But, if we assume no slipping, $v = r\omega$ and so:

$$20\,000 = 15 \times (v/0.300)^2 + 25v^2 + 14\,715$$

Hence $v = 5.25$ m/s. Using $v^2 = u^2 + 2as$ then $5.25^2 = 0 + 2a \times 30$ and so $a = 0.459$ m/s^2. Thus, using $v = u + at$ then $5.25 = 0 + 0.459t$ and so $t = 11.4$ s.

Revision

35   A particle of mass 8 kg is pulled 4 m up a smooth plane inclined at 30° to the horizontal. Determine the work done if the particle is pulled at a constant speed.

36   A car of mass 1500 kg travels along a horizontal road against a constant resistance to motion of 500 N at a speed of 40 m/s. Determine the rate at which the engine is working.

37   A wheel with a mass of 7 kg, radius of 400 mm and radius of gyration 300 mm rolls, without slipping, from rest down an incline at an elevation to the horizontal of 30°. What will be its angular velocity when it has rolled 5 m down the slope?

38   A uniform rod of length 3 m, mass 2 kg is pivoted about that end. Initially the rod is at rest in the horizontal position. Determine its angular velocity when it is released and has swung through 90°. Neglect any frictional torque.

39   A flywheel is in the form of a solid uniform disc of radius 600 mm and mass 290 kg and is rotating at 5 rev/s. Determine the tangential force which has to be applied to the edge of the wheel to bring it to rest in 10 s.

Problems

1   An object has an initial velocity of 3 m/s and is given a uniform acceleration of 2 m/s^2 for 6 s. What will be the resulting velocity?

2   An object initially at rest is given a uniform acceleration of 2 m/s^2 for 4 s. What will be the distance covered in that time?

3   An object with an initial velocity of 30 m/s is slowed down with a uniform acceleration of −4 m/s^2 until the velocity is 10 m/s. What will be the time taken?

4 An object, starting from rest, has a uniform acceleration of 2 m/s^2. What will be the distance covered in (a) the first four seconds, (b) the fourth second of the motion?

5 An object moves along a straight line from point A to point B. It has a velocity of 3 m/s at A and accelerates between A and B with a uniform acceleration of 0.5 m/s^2 to attain a velocity of 5 m/s^2 at B. What is the distance between A and B and the time taken for the object to move from A to B?

6 What initial velocity must an object have if it is to come to rest in 100 m as a result of a uniform retardation of 8 m/s^2?

7 A cam is designed so that its cam follower increases its velocity at a uniform rate from 3 m/s to 5 m/s when it moves vertically through a distance of 16 mm. Determine the time taken and the acceleration.

8 An object accelerates from rest with a uniform acceleration of 4 m/s^2 and then decelerates at 8 m/s^2 until at rest again, covering a total distance from rest-to-rest of 5000 m. What is the time taken?

9 A car accelerates along a straight road with uniform acceleration. When it passes point A it has a speed of 12 m/s and when it passes point B it has a speed of 32 m/s. Points A and B are 1100 m apart. Determine the time taken to move from A to B.

10 The velocity v of an object moving along a straight line path is related to the time t by $v = 4 + 2t$ m/s. Determine the acceleration.

11 The distance s travelled by an object moving along a straight line path is related to the time t by $s = 10t + 4t^2$. Determine, after 2 s, (a) the velocity, (b) the acceleration.

12 A stone is thrown vertically upwards from a point some distance above the ground level with an initial velocity of 20 m/s. If the stone hits the ground after 5 s, how far below the point of projection is the ground?

13 An object is thrown vertically upwards with an initial velocity of 14 m/s. Determine the height of the object above the point of projection after (a) 1 s, (b) 2 s.

14 Determine the resultant velocity of an object if it has velocities of 24 m/s in a northerly direction and 7 m/s in an easterly direction.

15 A raindrop falls in still air with a velocity of 3 m/s. What will be its velocity when there is a horizontal air current of 4 m/s?

16 Determine the components in northerly and easterly directions of a velocity of 2 m/s in a north-easterly direction.

17   What will be the greatest range on a horizontal plane for a projectile when the magnitude of the velocity of projection is 20 m/s?

18   Neglecting air resistance, what is the maximum range of a bullet from a rifle if it leaves the rifle with a velocity of 400 m/s.

19   A projectile has an initial velocity of 40 m/s at an elevation of 60° from the horizontal. Determine the speed of the projectile after 3 s.

20   A stone is thrown horizontally from the top of a tower with a velocity of 12 m/s. If the tower has a height of 10 m, determine how far from the foot of the tower the stone will land.

21   A bullet has an initial velocity of 600 m/s at an elevation of 25° to the horizontal. What will be its range?

22   A flywheel rotating at 210 rev/min is uniformly accelerated to 250 rev/min in 5 s. Determine the angular acceleration and the number of revolutions made by the flywheel in that time.

23   A flywheel rotating at 0.5 rev/s is uniformly accelerated to 1.0 rev/s in 10 s. Determine the angular acceleration and the number of revolutions made by the flywheel in that time.

24   A grinding wheel is rotating at 50 rev/s when the power is switched off. It takes 250 s to come to rest. What is the average angular retardation?

25   A wheel of diameter 350 mm rotates with an angular velocity of 6 rad/s. What is the speed of a point on its circumference?

26   A flywheel of diameter 360 mm increases its angular speed uniformly from 10.5 to 11.5 rev/s in 11 s. Determine (a) the angular acceleration of the wheel, (b) the linear acceleration of a point on the wheel rim.

27   A 160 mm diameter drive pulley is rotating at 400 rad/s and drives, via a belt drive, a second pulley of diameter 360 mm. Determine the uniform angular acceleration required for the drive pulley if the driven pulley is to be accelerated to an angular rotation of 200 rad/s after 6 s of acceleration.

28   A 200 mm radius pulley is rotating at 1200 rev/min and is used to drive, via a belt drive, a second pulley. Determine the radius required for the second pulley if it is to rotate a shaft at 800 rev/min.

29   A gear A with radius 50 mm is rotating at 30 rad/s and drives a second, intermeshed, gear B of radius 200 mm. If the gear B must rotate through 30 rad after 2 s, what is the angular acceleration needed for gear A?

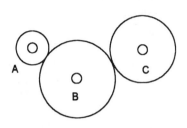

Figure 4.50   *Problem 30*

30 For the three gears shown in Figure 4.50, gear A has a radius of 60 mm, gear B a radius of 140 mm and gear C a radius of 120 mm. When gear A is rotating at 200 rev/min, what will be the angular velocity of gear C?

31 An object of mass 4 kg rests on a smooth plane which is inclined at 60° to the horizontal. The object is connected to one end of a light inextensible string which passes over a smooth fixed pulley at the top of the plane and from the vertical end of which another object of mass 2 kg is attached. Determine the acceleration of the objects and the tension in the string.

32 An object of mass 5 kg rests on a smooth horizontal table and is connected by a light inextensible string passing over a smooth fixed pulley at the edge of the table to a vertically suspended mass of 3 kg. Determine the acceleration of the objects and the tension in the string.

33 An object rests on a smooth horizontal table. On its left it is attached by a light inextensible string passing over a smooth fixed pulley on the edge of the table to a vertically suspended mass of 10 kg and on the left by another light inextensible string passing over a pulley at the edge of the table to a vertically suspended mass of 2 kg. Both the strings are attached to the object along the same horizontal line. Determine the mass of the object on the table if, when the system is released from rest, it accelerates at 2 m/s².

34 A car of mass M pulls a trailer of mass m along a level road. If the car engine exerts a forward force F and the tension in the tow bar is T, show that the acceleration of the car and trailer is $F/(M + m)$ and the tension in the tow bar is $F - Ma$.

35 A car of mass 900 kg pulls a trailer of mass 600 kg along a level road. The car experiences a resistance of 200 N and the trailer a resistance of 300 N. If the car exerts a forward force of 3000 N, determine the acceleration of the car and trailer and the tension in the tow bar.

36 An object of mass m rests on a smooth horizontal table and is connected to a freely hanging object of mass M by a light inextensible string passing over a smooth fixed pulley at the edge of the table. If the object on the table starts from rest a distance d from the pulley at the edge of the table, determine the time taken for it to reach the pulley.

37 A child of mass 25 kg sits on a toboggan of mass 15 kg on a snow slope inclined at 10° to the horizontal. If the toboggan slides down the slope with an acceleration of 1.2 m/s², what is the coefficient of friction?

38   An object of mass 9 kg is held, just on the point of moving, on a rough plane which is inclined at 30° to the horizontal by a force of 12 N acting parallel to the plane. What is the coefficient of friction?

39   An object of mass 5 kg is on a rough plane which has an angle of inclination to the horizontal which is gradually increased. When the angle reaches 13.5° the object begins to slide down the plane. What is the coefficient of friction? What would be the minimum force parallel to the plane which would be necessary to pull the block up the plane?

40   An object of mass m rests on a rough plane inclined at 30° to the horizontal and is attached, by means of a light inextensible string passing over a pulley at the top of the slope to a mass M which hangs vertically. The coefficient of friction is $1/\sqrt{3}$. Show that when the system is released that there will be an acceleration of $(M - m)g/(M + m)$.

41   A flywheel has a mass of 2 kg and a radius of gyration about an axis at right angles to its face and through its centre of 100 mm. Determine its moment of inertia about this axis.

42   A flywheel is a solid disc of mass 120 kg and radius 150 mm. Determine its moment of inertia about an axis at right angles to its plane and through its centre.

43   Determine the radius of gyration for an object having a moment of inertia of 0.5 kg m² and a mass of 2 kg.

44   Determine the moment of inertia about the rotational axis of the circular flywheel with the cross-section shown in Figure 4.51.

45   A dumb-bell consists of a slender rod of length 1.20 m and mass 10 kg with uniform spheres attached to the ends, each sphere having a mass of 45 kg and a radius of 100 mm. Determine the moment of inertia of the dumb-bell about an axis passing through the centre of the rod and at right angles to it.

46   A solid uniform disc has a mass of 120 kg and a radius of 150 mm. With the disc initially at rest, a constant torque of 10 N m is applied for 2 s. Determine the angular velocity of the disc after 2 s.

47   A wheel is mounted on a horizontal axle of radius 10 mm which is supported by bearings which can be assumed to be frictionless. A cord is wrapped round the axle and attached to a load of 5 kg. With the cord taut, when the load is released it falls through a distance of 1 m in 10 s. Determine the moment of inertia of the wheel and axle.

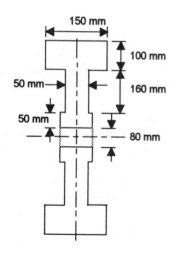

Figure 4.51   *Problem 44*

48 A solid, uniform, drum of mass 40 kg and radius 500 mm has a rope wrapped round its periphery. What will be the angular acceleration of the drum when a force of 200 N is applied to the rope.

49 A flywheel has a mass of 360 kg and a radius of gyration of 600 mm. If it is rotating at 10 rev/s, determine the uniform torque that has to be applied to bring it to rest in 30 s.

50 A uniform solid sphere of radius r rolls through one complete revolution from rest in rolling down a rough inclined plane at an elevation of θ from the horizontal. show that the resulting velocity of the centre of mass of the sphere is:

$$v = \sqrt{\frac{20\pi gr \sin \theta}{7}}$$

51 A uniform solid sphere rolls from rest a distance of 10 m down an inclined plane at an elevation of 30° from the horizontal. Determine the time taken.

52 A uniform bar of length 1 m has a mass of 10 kg and is pivoted about one end. Initially the bar is horizontal and at rest. Determine its angular velocity when it has rotated from this position through 60°. Neglect any frictional torque.

5 Mechanical oscillations

5.1 Introduction

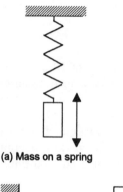

(a) Mass on a spring

(b) Cantilever

(c) Simple pendulum

Figure 5.1 *Examples of
mechanical oscillation systems*

This chapter is about mechanical oscillations; obvious examples of such oscillations are those of a mass suspended from the end of a vertical spring (Figure 5.1(a)), the oscillations of a loaded cantilever (Figure 5.1(b)) and the angular oscillations of a simple pendulum (Figure 5.1(c)). Basically we can think of all mechanical objects as having mass and elasticity and being capable of oscillations. Such oscillations often can be detrimental and result in the failure of machine parts, excessive wear or just undesirable noise.

5.1.1 Basic terms

The following are basic terms used to describe oscillations:

1 *Periodic time, cycle and frequency*
   Oscillatory motions are periodic, i.e. they keep on repeating themselves after equal intervals of time (Figure 5.2). The time between repetitions is called the *periodic time* and the oscillatory motion occurring within one period is called a *cycle*. The number of cycles per second is called the *frequency*. Thus, since one cycle occurs in the periodic time T, the frequency is $1/T$.

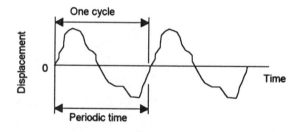

Figure 5.2 *A periodic oscillation*

2 *Free oscillations and natural frequency*
   The term *free oscillation* is used when an elastic system oscillates under the action of forces inherent in the system itself with there being no externally applied forces. For example, with the spring system in Figure 5.1(a) when it is given an initial deflection to start the oscillation and then left to freely oscillate. The system will oscillate at, what is termed, a *natural frequency*, such frequencies being determined by the properties of the system.

3   *Forced oscillations*
Oscillations which take place under the effect of externally applied periodic forces are called *forced oscillations*. Thus if the support of the spring system in Figure 5.1(a) was itself oscillating it would give forced oscillations of the spring system.

4   *Damping*
Oscillating mechanical systems are all subject to damping due to energy being dissipated by friction and other resistances. With free oscillations, since no energy is externally supplied to the system when oscillating then the effect of energy being dissipated by damping is for the oscillations to die away with time. Thus if the simple pendulum in Figure 5.1(c) is set into free oscillation, the oscillation will gradually die away with time with the amplitude of the oscillation becoming progressively smaller.

5.2 Simple harmonic motion

The simplest form of periodic motion is *simple harmonic motion*. Consider a basic mechanical system of mass which when deflected from its rest position is restored to it by forces arising from elasticity in the system. Figure 5.3 shows such a system as a trolley tethered between two supports by springs. A trolley is considered for the mass in order to effectively eliminate frictional effects.

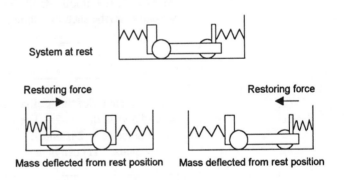

System at rest

Restoring force

Restoring force

Mass deflected from rest position   Mass deflected from rest position

Figure 5.3   *A basic mechanical system*

When the trolley, the mass of the system, is pulled to one side then one of the springs is compressed and the other stretched and this has the effect of providing a restoring force which is directed in such a direction as to endeavour to restore the trolley back to its original position. If the trolley is released from this deflected position, the restoring force causes the trolley to move back towards its original rest position and overshoot that position. The restoring force then reverses its direction to still be directed towards the rest position and so oscillations occur. If the displacement from the rest position is measured as a function of time then the result is as shown in Figure 5.4, the displacement variation with time being described by a cosine graph.

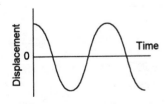

Figure 5.4   *Graph of the displacement with time*

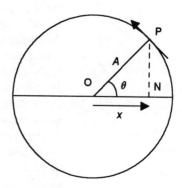

Figure 5.5 *Horizontal displacement for a point moving in a circular path with a constant angular velocity*

Such a type of displacement variation with time can be produced by the horizontal displacement x from the centre of a point P rotating in a circular path with a constant angular velocity ω (Figure 5.5), i.e.

$x = A \cos \theta$

where A is the amplitude of the oscillation. Since we have $\theta = \omega t$, then we can describe such oscillations by:

$$x = A \cos \omega t \qquad\qquad [1]$$

The frequency f of the oscillations is the number of cycles completed per second and is thus $\omega/2\pi$. The term *angular frequency* is sometimes used for ω since $\omega = 2\pi f$ and it is thus just the frequency multiplied by 2π. Thus equation [1] can be written as:

$$x = A \cos 2\pi f t \qquad\qquad [2]$$

The linear velocity v at some instant is the rate of change of displacement dx/dt and thus, differentiating equation [1] gives:

$$v = -A\omega \sin \omega t \qquad\qquad [3]$$

In Figure 5.5 we have $\sin \theta = PN/A$, and since $PN^2 = A^2 - x^2$ then $A \sin \theta = \sqrt{(A^2 - x^2)}$ and so equation [3] can be written as:

$$v = -\omega\sqrt{A^2 - x^2} \qquad\qquad [4]$$

The maximum velocity is when $x = 0$, i.e. as the mass passes through its rest position, and is:

$$\text{maximum velocity} = -\omega A \qquad [5]$$

The linear acceleration a at an instant is the rate of change of velocity dv/dt and thus differentiating equation [3] gives:

$$a = -\omega^2 A \cos \omega t$$

This can be written as:

$$a = -\omega^2 x \qquad [6]$$

The acceleration has a maximum value when x equals the maximum displacement A:

$$\text{maximum acceleration} = -\omega^2 A \qquad [7]$$

The restoring force $F = ma$ and is thus given by equation [6] as:

$$F = -m\omega^2 x \qquad [8]$$

The minus sign indicates that the direction of this restoring force is always in the opposite direction to that for which x increases. Simple harmonic motion is thus defined as:

Simple harmonic motion (SHM) is said to occur when the motion is under the action of a restoring force which is always directed to a fixed point and has a magnitude which is proportional to the displacement from that point.

We can write, using equation [8], the angular frequency as:

$$\omega = \sqrt{\frac{F}{mx}}$$

and thus, since the periodic time $T = 1/f = 2\pi/\omega$:

$$T = 2\pi \sqrt{\frac{\text{mass}}{\text{force per unit displacement}}} \qquad [9]$$

or:

$$f = \frac{1}{2\pi} \sqrt{\frac{\text{force per unit displacement}}{\text{mass}}} \qquad [10]$$

The larger the force needed to produce unit displacement the higher the frequency. Thus a high 'stiffness' mechanical system with its large force per unit displacement will have a high frequency of oscillation.

Example

An object moving with simple harmonic motion has an amplitude of 1.2 m and a periodic time of 3 s. Determine the maximum velocity and maximum acceleration and state at what points in the oscillation they occur.

The angular frequency $\omega = 2\pi f = 2\pi/T = 2\pi/3 = 2.09$ rad/s. Thus, using equation [5]:

$$\text{maximum velocity} = -\omega A = -2.09 \times 1.2 = -2.51 \text{ m/s}$$

The maximum velocity occurs when the displacement from the rest position is zero. The maximum acceleration is given by equation [7] as:

$$\text{maximum acceleration} = -\omega^2 A = -2.09^2 \times 1.2 = -5.24 \text{ m/s}^2$$

The maximum acceleration occurs when the displacement from the rest position is a maximum.

Example

An object moves with simple harmonic motion and has an amplitude of 500 mm and a frequency of 4 Hz. Determine the velocity and acceleration of the object when it is 200 mm from its rest position.

The angular frequency $\omega = 2\pi f = 2\pi \times 4 = 25.1$ rad/s. The velocity is thus given by equation [4] as:

$$v = -\omega \sqrt{A^2 - x^2} = -25.1\sqrt{0.5^2 - 0.2^2} = -11.5 \text{ m/s}$$

The acceleration is given by equation [6] as:

$$a = -\omega^2 x = -25.1^2 \times 0.2 = -126 \text{ m/s}$$

Revision

1   An object is oscillating with simple harmonic motion of amplitude 40 mm and frequency 40 Hz. Determine the velocity and the acceleration when the object is at the maximum displacement from the centre and at the centre of its oscillation.

2   The velocity of an object which is oscillating with simple harmonic motion is 4 m/s when it is 3 m from the central rest position and 3 m/s when it is 4 m. Determine the velocity of the object as it passes through the central rest position.

3   An object of mass 0.1 kg oscillates with simple harmonic motion of frequency 15 Hz. Determine the restoring force acting on the object when it is at a displacement of 30 mm.

5.2.1 Energy of simple harmonic motion

The velocity of an object when oscillating with simple harmonic motion and at a displacement x from its central rest position is given by equation [4] as:

$$v = -\omega\sqrt{A^2 - x^2}$$

Thus the kinetic energy is:

kinetic energy $= \tfrac{1}{2}mv^2 = \tfrac{1}{2}m\omega^2(A^2 - x^2)$ [11]

Because the restoring force is proportional to the displacement, the work done to move the object from its central rest position to a displacement x is given by the average force acting over that displacement multiplied by the displacement. The force is zero at the central position and given by equation [8] as $m\omega^2 x$ at displacement x. Thus:

work done $= \tfrac{1}{2}m\omega^2 x^2$

Thus the potential energy of the object when displaced by x is:

potential energy $= \tfrac{1}{2}m\omega^2 x^2$ [12]

The total energy at this displacement is the sum of the potential and kinetic energies (equations [11] and [12]) and thus is:

energy $= m\omega^2 A^2$ [13]

The total energy is thus constant at all displacements, depending only on the amplitude of the oscillation. What varies at different displacements is the fraction of the energy that is kinetic energy and the fraction that is potential energy. As an object oscillates there is a continual changing of potential energy to kinetic energy and vice versa with, in the absence of losses or inputs to the system, i.e. damping or forcing, the sum remaining constant.

Example

Determine the energy of an object of mass 0.5 kg oscillating with simple harmonic motion of amplitude 100 mm and frequency 5 Hz.

The angular frequency $\omega = 2\pi f = 2\pi \times 5 = 31.4$ rad/s. Thus, using equation [13]:

$$\text{energy} = m\omega^2 A^2 = 0.5 \times 31.4^2 \times 0.1^2 = 4.93 \text{ J}$$

Revision

4   An object of mass 0.1 kg oscillates with simple harmonic motion of frequency 15 Hz and amplitude 50 mm. Determine the total energy of the object.

5.3 Undamped oscillations

The following are derivations of the natural frequencies of a number of mechanical systems executing oscillations when damping is assumed to be negligible. The derivations are based on the technique of determining how the restoring force varies with displacement and then, provided the restoring force is proportional to the displacement and directed towards the rest position, using equation [9] to obtain the frequency.

5.3.1 Mass on a spring

Consider a mass suspended from a vertical spring (Figure 5.6), the mass of the spring being assumed to be negligible. It is made to oscillate in the vertical direction by the mass being pulled down, so extending the spring, and then the mass is released. The spring then exerts a restoring force on the mass. Assuming that the spring obeys Hooke's law, then the restoring force F is proportional to the displacement x of the end of the spring from its rest position. The force is always directed towards the rest position and thus we can write:

$$F = -kx$$

Figure 5.6   *Mass on a spring*

where k is the spring stiffness. The motion is simple harmonic because F is proportional to $-x$. The magnitude of the force per unit displacement is k and thus, using equation [9], i.e.

$$f = \frac{1}{2\pi} \sqrt{\frac{\text{force per unit displacement}}{\text{mass}}}$$

then:

$$f = \frac{1}{2\pi} \sqrt{\frac{k}{m}} \qquad [14]$$

For two springs in series (Figure 5.7(a)), one having a stiffness k_1 and the other a stiffness k_2, if a force F is applied then the same force must be applied to each spring and so the series arrangement will stretch by $x = x_1 + x_2$, where $x_1 = F/k_1$ and $x_2 = F/k_2$. Thus the overall system will have a stiffness k of:

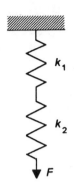

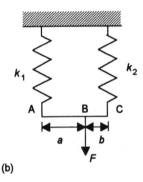

$$k = \frac{F}{x} = \frac{1}{\frac{1}{k_1} + \frac{1}{k_2}} = \frac{k_1 k_2}{k_1 + k_2} \qquad [15]$$

(a)

For two springs in parallel (Figure 5.7(b)), one having a stiffness k_1 and the other a stiffness k_2, if a force F is applied at point B then if we take moments about A we have $Fa = F_2(a + b)$, where F_2 is the force acting on the spring with stiffness k_2 and if we take moments about C we have $Fb = F_1(a + b)$, where F_1 is the force acting on the spring with stiffness k_1. The deflection of point A is thus $x_1 = F_1/k_1 = Fb/k_1(a + b)$ and the deflection of point C is $x_2 = F_2/k_2 = Fa/k_2(a + b)$. As indicated by Figure 5.8, the deflection x of point B is:

$$x = x_1 + \frac{a}{a+b}(x_2 - x_1)$$

$$= \frac{Fb}{k_1(a+b)} + \frac{a}{a+b}\left(\frac{Fa}{k_2(a+b)} - \frac{Fb}{k_1(a+b)}\right)$$

$$= \frac{F}{(a+b)^2}\left(\frac{a^2}{k_2} + \frac{b^2}{k_1}\right)$$

(b)

Figure 5.7 *Springs in (a) series, (b) parallel*

and thus the effective stiffness k of the spring arrangement is:

$$k = \frac{F}{x} = \frac{(a+b)^2}{\left(\dfrac{a^2}{k_2} + \dfrac{b^2}{k_1}\right)} \qquad [16]$$

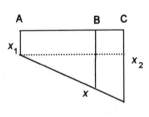

Figure 5.8 *Deflections*

If $k_1 = k_2$ and $a = b$, then $k = 2k_1$.

Example

A spring supports a carrier of mass 2 kg and when a 10 kg mass is placed on the carrier the spring extends by 50 mm. The carrier and load are then pulled down a further 75 mm and released. Determine the frequency of the oscillations for vertical oscillations.

A force of $10g$ N causes an extension of 50 mm and thus, assuming the spring obeys Hooke's law, the stiffness $k = 10g/0.050 = 1962$ N/m. The frequency of oscillation is given by equation [14] as:

$$f = \frac{1}{2\pi}\sqrt{\frac{k}{m}} = \frac{1}{2\pi}\sqrt{\frac{1962}{12}} = 2.04 \text{ Hz}$$

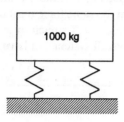

Figure 5.9   *Example*

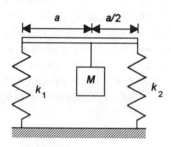

Figure 5.10   *Revision problem 7*

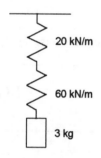

Figure 5.11   *Revision problem 8*

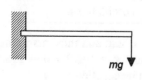

Figure 5.12   *Cantilever*

Example

A machine of mass 1000 kg is mounted centrally on two rubber pads, its centre of mass being central (Figure 5.9). Each pad has a stiffness of 500 kN/m. Determine the natural frequency of oscillation of the system.

The total stiffness of the supports is (see note under equation [16]) $2 \times 500 = 1000$ kN/m. Thus the frequency is given by equation [14] as:

$$f = \frac{1}{2\pi} \sqrt{\frac{k}{m}} = \frac{1}{2\pi} \sqrt{\frac{1000 \times 1000}{1000}} = 5.0 \text{ Hz}$$

Revision

5   A mass of 100 kg is suspended from a vertical spring and stretches it by 50 mm. Determine the natural frequency of the system when it is set oscillating vertically.

6   A mass of 10 kg is suspended from a vertical spring with a spring constant of 2 kN/m. Determine the natural frequency of the system when it is set oscillating vertically.

7   Determine the natural frequency for vertical oscillations of the system shown in Figure 5.10.

8   Determine the natural frequency for vertical oscillations of the system shown in Figure 5.11, it being a mass of 3 kg suspended by two series connected springs of stiffness 20 kN/m and 60 kN/m.

5.3.2 Cantilever

Consider a cantilever of negligible mass with a point load of mass m at its free end (Figure 5.12), the load resulting in a vertical deflection d from the horizontal. Since the deflection is proportional to the load we have a stiffness $k = mg/d$. If the cantilever is now pulled down a further distance x at the free end, then the restoring force is $F = -kx = -mgx/d$. The restoring force is proportional to the displacement and thus when the cantilever is released it performs oscillations with simple harmonic motion. Hence, using equation [9], i.e.

$$f = \frac{1}{2\pi} \sqrt{\frac{\text{force per unit displacement}}{\text{mass}}}$$

then:

$$f = \frac{1}{2\pi} \sqrt{\frac{mg/d}{m}} = \frac{1}{2\pi} \sqrt{\frac{g}{d}} \qquad [17]$$

Example

A machine is placed on a horizontal steel girder which is fixed at both its ends. As a result of placing the machine on the girder it deflects by 2 mm at the point where the machine is mounted. Estimate the natural frequency of the transverse oscillations that can occur if the mass of the girder is neglected and the deflection is proportional to the load.

As with the cantilever, the deflection d is proportional to the load and so we have a stiffness $k = mg/d$. This is the same as the theory derived above for a load at the end of a cantilever and so we can use the same equation, namely equation [17], for the estimate. Thus:

$$f = \frac{1}{2\pi}\sqrt{\frac{g}{d}} = \frac{1}{2\pi}\sqrt{\frac{9.81}{0.002}} = 11.1\ \text{Hz}$$

Revision

9   What will be the frequency of transverse oscillations of the cantilever when a mass of 2 kg is attached to the free end? The stiffness of the cantilever is such that when a force of 9 N is applied to the free end of a cantilever a deflection of 75 mm is produced.

5.3.3 Simple pendulum

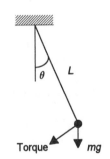

Figure 5.13   *Pendulum*

Consider a simple pendulum in which the mass m of the pendulum acts at a distance L from the point of suspension (Figure 5.13). Note the use of the term 'simple', this is because we will make the assumption that all the mass of the pendulum is concentrated in a small bob on the end of a string of negligible mass; if these assumptions cannot be made the pendulum is termed compound and the analysis of the motion of the pendulum has to be modified from that which follows. When the pendulum is pulled through a small angle θ from the vertical, there is a resultant torque acting on the pendulum bob which arises from the component of the weight mg which is tangential to the arc of motion of the bob, i.e. $mg \sin \theta$. The torque is thus $-mgL \sin \theta$. Since we are only considering small angles $\sin \theta \approx \theta$ and thus the torque is approximately $-mgL\theta$. This torque produces an angular acceleration a, where:

$$\text{torque} = -mgL\theta = Ia$$

where I is the moment of inertia of the system. Since the pendulum is a mass concentrated at a distance L from the axis of rotation, the moment of inertia is mL^2. Thus:

$$-mgL\theta = mL^2a$$

The angular acceleration is thus proportional to the angular displacement $-\theta$. This compares with the statement already used for simple harmonic motion where the linear acceleration is proportional to the displacement and we consider angular simple harmonic motion to be occurring. Note that we only obtain simple harmonic motion when we can make the approximation $\sin \theta \approx \theta$. The above equation gives:

$$\text{angular acceleration} = -\frac{g}{L}\theta \qquad [18]$$

This is the angular equivalent of equation [6], i.e.

$$a = -\omega^2\theta \qquad [19]$$

and thus, since $\omega = 2\pi f$:

$$f = \frac{1}{2\pi}\sqrt{\frac{g}{L}} \qquad [20]$$

Note that equation [19] gives $T = Ia = -I\omega^2\theta$ and so the equivalent of equation [9] for angular simple harmonic motion as:

$$f = \frac{1}{2\pi}\sqrt{\frac{\text{torque/unit angular displacement}}{\text{moment of inertia}}} \qquad [21]$$

Example

How does the periodic time of a simple pendulum with a length of 1.21 m compare with that for a pendulum with length 1.00 m?

Using equation [20], since the periodic time is the reciprocal of the frequency, we have:

$$\text{periodic time} \propto \sqrt{L}$$

and thus:

$$\text{ratio of the periodic times} = \sqrt{\frac{1.21}{1.00}} = 1.10$$

Revision

10   A pendulum which has a periodic time of 1.000 s when its support is stationary is placed in a lift which ascends with an acceleration of 0.02 m/s. What effect will this have on the periodic time?

5.3.4 Torsional oscillations

Figure 5.14   *Torsional oscillations*

Consider a disc, with moment of inertia I about an axis at right angles to its plane and through its centre, which is mounted on a central axial slender shaft (Figure 5.14). When the disk is rotated through some angle

θ, the shaft is twisted. The restoring torque T produced by this twist is proportional to θ (see Chapter 3 and equation [10], i.e. $T = G\theta J/L$ with G = shear modulus, J = polar second moment of area and L = length), thus:

restoring torque $= -k\theta$

where k is the torque per radian twist or torsional stiffness (see note above, it is GJ/L). When the system is released and allowed to oscillate then:

restoring torque $= I\alpha = -k\theta$

The oscillation is thus angular simple harmonic and so, using equation [21]:

$$f = \frac{1}{2\pi}\sqrt{\frac{k}{I}} \qquad\qquad\qquad [22]$$

Example

A flywheel of mass 30 kg and radius of gyration 150 mm about its central axis is attached centrally to one end of a slender shaft, the other end being effectively clamped. The shaft has a torsional stiffness of 1.6 kN m/rad. Determine its frequency of oscillations when the flywheel is given an initial angular displacement and then released.

The moment of inertia of the flywheel is $I = mk^2 = 30 \times 0.150^2 = 0.675$ kg m^2. Thus, using equation [22]:

$$f = \frac{1}{2\pi}\sqrt{\frac{1600}{0.675}} = 7.75 \text{ Hz}$$

Revision

Figure 5.15   *Revision problem 11*

11 A rectangular plate 200 mm by 300 mm and having a mass of 10 kg is fixed by its centre to the end of a slender rod having a torsional stiffness of 1.5 N m/rad (Figure 5.15). Determine the natural frequency of oscillation when the plate is given an angular displacement and then released.

5.4 Damped oscillations

So far in this chapter in the discussion of oscillations, it has been assumed that the only force acting on an object performing oscillations is the restoring force. However, there are other forces, since if there was only the restoring force an object would continue oscillating for ever. Thus if the oscillations of a mass suspended from a fixed support are considered; when the mass is pulled down and then released it oscillates with the amplitude of each successive oscillation becoming smaller

(Figure 5.16), eventually stopping as a result of frictional forces. When damping occurs there is a loss of energy from the oscillating system as work is done against such forces as friction or air resistance. Since the total energy of an oscillating system is (equation [13]):

$$\text{energy} = m\omega^2 A^2$$

then diminishing this energy as a result of energy loss means the amplitude decreases. The oscillation is said to be *damped*, the amount of damping determining how rapidly the amplitude diminishes.

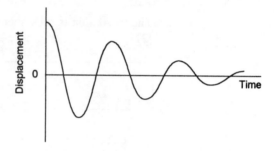

Figure 5.16   *Damping reducing the amplitude of the oscillation*

Figure 5.16 shows how the displacement varies with time with moderate damping. If the damping is increased there comes a point at which the oscillations just fail to materialise and the displacement becomes of the form shown in Figure 5.17. Such damping is termed *critical damping*. Critical damping is defined as the value of damping needed for there to be no oscillations and the displacement reach zero in the minimum time. If the damping is increased beyond this, the time to reach zero displacement is increased; the system is said to be *over damped*. The system with less damping than critical damping gives oscillations and is said to be *under damped*.

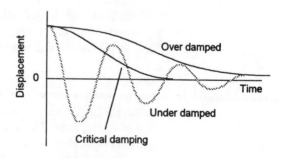

Figure 5.17   *Effect of different degrees of damping*

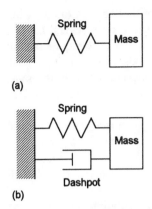

(a)

(b)

Dashpot

Figure 5.18   *Models for (a) undamped, (b) damped systems*

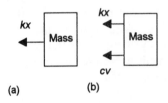

(a)   (b)

Figure 5.19   *Free-body diagrams for (a) undamped, (b) damped systems*

Car shock absorbers have damping mechanisms deliberately introduced so that the vibrations of the suspension relative to the car frame are rapidly damped out. Without this, passengers in their seat would oscillate for many oscillations every time the car hits a bump in the road. With an instrument pointer of a moving galvanometer, when a current occurs the coil of the instrument is given a deflection and can start oscillating. Such oscillations are generally undesirable since we want the instrument pointer to quickly settle down to the steady-state reading. Thus damping is introduced. This typically involves using a metal former for the galvanometer coil, its motion in the magnetic field of the instrument resulting in eddy currents being produced which oppose the coil motion.

A basic model for an undamped oscillating system is one which has just mass and elasticity (Figure 5.18(a)). We can modify the model to include damping by introducing a dashpot (Figure 5.18(b)). When the piston of the dashpot moves, air has to either escape from or get into the trapped air space. The result is a resistance to motion which depends on the velocity with which the piston moves, the faster it moves the greater the resistance to motion. With the undamped system, the restoring force acting on the mass is $-kx$ and the free-body diagram is as shown in Figure 5.19(a). Thus, for the acceleration a of the mass m, we have:

$$-kx = ma$$

With the damped system, there is a damping force acting on the mass which is typically proportional to the velocity v (this is normally the case where damping is due to air resistance, the viscosity of a liquid or electromagnetic induction) and the free-body diagram is as shown in Figure 5.19(b). Thus we have:

$$-kx - cv = ma$$

where c is the damping constant.

Example

A mass oscillating on the end of a spring is damped so that it loses 16/25 of its energy in each cycle. How does the amplitude vary with time?

Equation [13] gives the energy as $m\omega^2A^2$. Thus the amplitude is proportional to the square root of the energy. The energy in successive cycles is E and $9/25E$. Thus the ratio of the amplitude in successive cycles is $\sqrt{(9/25)} = 3/5$. Each successive amplitude is three-fifths that of the previous oscillation.

5.5 Forced oscillations

If you give a child on a swing an initial push, the oscillations will usually have a periodic time of about two or three seconds and will gradually diminish in amplitude unless further pushes are given. To build up large

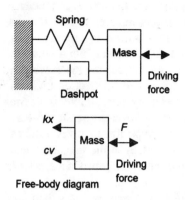

Figure 5.20   *Model for forced oscillation system*

amplitude oscillations it is necessary for the pushes applied to the swing to be at just the right time intervals. If we tried to push several times a second then the amplitude of the oscillation would be quite small; similarly if we pushed just once every ten seconds. However, it we push each time the swing is at its maximum displacement, i.e. with the same periodic time as that of the swing, then very large amplitude oscillations can be built up.

The above represents a situation where an object with a natural frequency is supplied with an externally applied periodic force. The resulting oscillations are said to be *forced oscillations*. The frequency of the applied external force is called the *driver frequency*. Figure 5.20 shows the basic model for such a system

Figure 5.21 shows how the amplitude of the oscillation depends on the driver frequency for different values of damping in the system. When the largest amplitude oscillations occur the system is then said to be in *resonance*. The less the damping the larger the amplitude at the resonance frequency and the 'sharper' the graph, i.e. the large response over a narrow range of frequencies. With more damping there is a smaller response and the graph is less 'sharp'. With very low damping, resonance occurs when the driver frequency is equal to the natural frequency of the oscillator. At resonance with low damping, relatively small forces can cause very large amplitude oscillations.

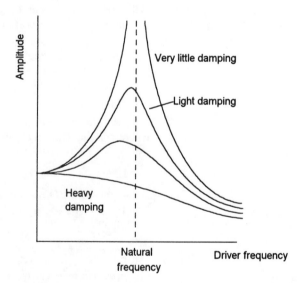

Figure 5.21   *The amplitude of oscillations as a function of driver frequency and damping*

Resonance is a phenomenon that occurs in many situations. For example, if the wheel of a car is out of balance it can oscillate like a mass on a spring, i.e. the car suspension, and so has a natural frequency. At certain speeds there might be a forcing frequency applied which is close enough to the natural frequency to cause relatively large amplitude oscillations and the driver of the car can more readily perceive the oscillations, even though the road surface over which the car was being driven was not particularly rough. Aeroplanes are tested, by shaking them at a wide range of frequencies, to determine their resonant frequencies and so ensure that they are unlikely to occur when the plane is in service. If they did occur, the resulting large amplitude oscillations could result in damage to the plane. Soldiers marching in step over a bridge apply a forcing frequency and, though the force applied is small, if it is near to the natural frequency it can result in large amplitude oscillations which destroy the bridge. This happened in 1831 in Manchester when 60 marching soldiers broke the Brough suspension bridge over the River Irwell. In 1850 a French infantry battalion marched in step across the Angers suspension bridge and the resulting oscillations destroyed it with, as a consequence, the deaths of 226 of the soldiers. Soldiers marching across bridges have to break step to avoid the possibility of such an event occurring.

Problems

1   An object oscillates with simple harmonic motion of frequency 4 Hz and amplitude 150 mm. Determine the maximum velocity of the object.

2   An object oscillates with simple harmonic motion. At a displacement of 1 m from its central rest position the acceleration is 200 m/s^2. Determine the frequency of the oscillation.

3   An object oscillates with simple harmonic motion of frequency 4 Hz and amplitude 150 mm. Determine the velocity and acceleration of the particle at a displacement of 90 mm.

4   An object oscillates with simple harmonic motion. Determine the periodic time if the acceleration is 4 m/s^2 when the displacement from the centre of the oscillation is 2 m.

5   An object oscillates with simple harmonic motion and has a periodic time of 8 s and amplitude 4 m. Determine the maximum velocity and the velocity when the object is 2 m from its central rest position.

6   An object of mass 0.2 kg oscillates with simple harmonic motion of frequency 10 Hz and amplitude 70 mm. Determine the energy of the oscillating object.

7   A mass of 5 kg is suspended from a fixed support by a vertical spring. If the natural frequency of oscillation is 2 Hz for vertical oscillations, what is the stiffness of the spring?

8   Determine the natural frequency of oscillation for the system shown in Figure 5.22 of a mass of 20 kg suspended by two identical symmetrically spaced springs, each having a stiffness of 500 N/m.

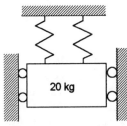

Figure 5.22   *Problem 8*

20 kg

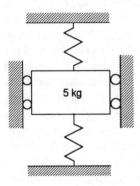

Figure 5.23   *Problem 9*

The arrangement is constrained so that only vertical oscillations can occur.

9   Determine the natural frequency of oscillation for the system shown in Figure 5.23 of a mass of 5 kg suspended by two identical springs, each having a stiffness of 500 N/m. The arrangement is constrained so that only vertical oscillations can occur.

10   A horizontal platform of mass 2 kg rests symmetrically on four vertical springs, each of stiffness 15 kN/m. Determine the natural frequency for vertical oscillations.

11   When four people with a combined mass of 300 kg get into a car, the springs are observed to compress by 50 mm. If the total load supported by the springs is 900 kg, determine the natural frequency of the system for vertical oscillations.

12   A simple pendulum has a periodic time of 1.000 s where the acceleration due to gravity is 9.81 m/s². What will be its periodic time when it is taken to a place where the acceleration due to gravity is 9.85 m/s²?

13   A flywheel of moment of inertia 5 kg m² about an axis perpendicular to its face and through its centre is attached centrally to one end of a slender shaft, the other end being effectively clamped. The shaft has a torsional stiffness of 500 N m/rad. Determine its frequency of oscillations when the flywheel is given angular displacement and then released.

14   A disc with a moment of inertia of 0.06 kg m² about an axis through its centre and at right angles to its face is attached at its centre to the lower end of a vertical rod, the other end of the rod being rigidly fixed. The rod has a uniform circular cross-section of radius 4.5 mm, a length of 1.75 m and modulus of rigidity of 80 GPa. When the disc is given an angular displacement and released, what will be the frequency of the oscillations?

15   A U-tube is partially filled with a liquid so that the total length of the liquid in the tube is L. When the liquid is set oscillating in the tube, show that the frequency of the oscillations is given by:

$$f = \frac{1}{2\pi}\sqrt{\frac{L}{2g}}$$

16   A cylinder with a uniform cross-section is floating vertically in a liquid with a depth h immersed. When the cylinder is pushed down a little way and released, oscillations occur. Show that the frequency of the oscillations is given by:

$$f = \frac{1}{2\pi}\sqrt{\frac{g}{h}}$$

17   A small coin rests on a horizontal platform which is oscillating with simple harmonic motion of frequency 2.5 Hz. What is the maximum amplitude of this motion which will allow the coin to remain in contact with the platform at all times?

18   Figure 5.3 shows a trolley of mass m on a horizontal table and tethered by springs at each end to fixed supports. For the springs initially taut and with identical spring stiffness k, and neglecting frictional effects, determine the frequency of the oscillation when the trolley oscillates back-and-forth along the line of the springs.

19   A mountaineer of mass 80 kg is on the end of a rope of length 35 m when he loses his hold and becomes suspended by the rope. If the rope stretches by 1.6 m under his weight, what will be the periodic time of his oscillations if (a) they are vertical oscillations, (b) he swings like a pendulum bob?

20   A mass oscillating on the end of a spring is damped so that it loses half of its energy in each cycle. How does the amplitude vary with time?

21   An oscillating system has a damping mechanism which dissipates 36% of the oscillator energy each complete oscillation. If the initial amplitude was 30 mm, what will be the amplitude after 1, 2 and 3 oscillations?

6 Heat transfer

6.1 Introduction   Energy transfer takes place by work or heat. *Work* is defined as energy transferred from one body to another because the point of application of a force moves through a distance. *Heat* is defined as energy transferred from one body to another because of a difference in temperature between them. Heat transfer processes are described by equations which relate the energy transferred in unit time to the temperatures and physical properties of the bodies involved in the transfer, e.g. the areas involved. There are three basic modes of heat transfer: *conduction, convection* and *radiation*. These may occur separately or simultaneously. Separate equations can be written for each mode of heat transfer and generally, when heat transfer is occurring by more than one mode, we can just consider each mode separately and then sum the results to give the total heat transfer.

There are two basic methods of transferring heat using matter as the transfer medium: conduction and convection. *Conduction* can basically be considered as being the mode of transfer when energy is just passed from one atom or molecule to another through the material. If you think of a crowd of people, the analogue of conduction is that energy is transferred through the crowd by one person jostling the next who in turn jostles the next one and so on, the jostling thus being passed through the crowd. An example of conduction is when one end of a metal bar is heated and, with the other end cool, heat energy is transferred through the bar from the hot end to the cool end. *Convection* occurs as a result of the gross motion of parts of the material. The analogue of convection with the crowd is that people start moving through the crowd and so a disturbance in one region of the crowd is transferred to another region by people moving to that region. An example of convection is when water is heated in a pan. The hot water at the bottom of the pan rises through the cooler water and so transfers heat from the hotter part of the water to the cooler part.

There is one method of transferrring heat which does not require matter for the trasnfer medium: radiation. *Radiation* is the transfer of heat energy by electromagnetic waves and, unlike conduction and convection, does not require matter as a medium for transfer. We are all familiar with the transfer of heat energy from the sun to the earth through the vacuum of space.

This chapter is a discussion of energy transfer by the three basic modes of heat transfer; Chapter 7 is also a discussion of energy transfer but restricted to that involved in the motion of fluids.

6.2 Conduction

Conduction is the mode of heat transfer that occurs as a result of the transfer of the internal energy of motion of constituent molecules and atoms from one to another though a material. It occurs by virtue of a temperature difference between different parts of the material. Conduction occurs in solids where it might be the sole method of heat transfer. It can also occur in liquids and gases but is then often occurring with other modes of heat transfer. With liquids there is generally also convection occurring and in the case of gases there is likely to be both convection and radiation.

The heat transfer rate by conduction is proportional to the temperature gradient in the direction of the heat flow and to the area of the material perpendicular to the direction of the heat flow. This is known as the *Fourier low of heat conduction*. It can be expressed as:

$$\text{heat transfer rate per unit area } q = -k\frac{\mathrm{d}T}{\mathrm{d}x} \qquad [1]$$

where $\mathrm{d}T/\mathrm{d}x$ is the temperature gradient and k a constant called the *thermal conductivity*. Because the temperature decreases in the direction of the heat flow, the equation includes a minus sign so that a negative temperature gradient gives a positive heat flow. Typical values of thermal conductivity are given in Table 6.1. Metals have much higher thermal conductivities than non-metals.

Table 6.1 Typical values of thermal conductivities

Material	k W/m K	Material	k W/m K
Copper	380	Fibreboard	0.11
Mild steel	54	Cork	0.04
Concrete	1.4	Glass wool blanket	0.04
Common brick	1.2	Expanded polystyrene	0.03
Softwood	0.13	Air	0.029

6.2.1 Heat conduction through plane layers

Consider heat conduction through a plane wall for which heat transfer in a right-angles direction is negligible (Figure 6.1). If the wall surfaces are at temperatures T_1 and T_2 and the wall thickness is x, the temperature gradient in the direction of the heat flow is $(T_1 - T_2)/x$ and so:

$$\text{heat transfer rate per unit area } q = -k\frac{T_1 - T_2}{x} \qquad [2]$$

For a plane wall of surface area A, the rate of heat transfer by conduction $Q = qA$ and thus:

$$\text{heat transfer rate } Q = -kA\frac{T_1 - T_2}{x_,} \qquad [3]$$

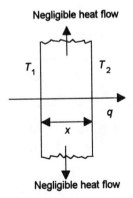

Negligible heat flow

T_1 T_2

q

x

Negligible heat flow

Figure 6.1 *Heat conduction through a plane wall*

Example

Calculate the rate of heat loss by conduction through the wall of a large industrial refrigerator if the wall has an area of 6 m² and consists of a layer of cork 200 mm thick sandwiched between metal sheets. The temperatures of the inner and outer surfaces of the cork are –15°C and 20°C. The cork has a thermal conductivity of 0.04 W/m K.

Assuming that we can regard the cork as a plane surface with all the heat transfer being at right angles to the wall, then equation [3] gives:

$$\text{heat transfer rate} = -0.04 \times 6 \times \frac{-15 - 20}{0.200} = 42 \text{ W}$$

Revision

1   Calculate the rate of heat loss by conduction though a single brick wall if the wall has an area of 20 m² and a thickness of 100 mm. The temperatures of the inner and outer surfaces of the wall are 20°C and 5°C. The brick has a thermal conductivity of 1.2 W/m K.

2   Determine the heat transferred by conduction per square metre of surface area through a wall of thickness 300 mm if one side of the wall is at a temperature of 230°C and the other at 15°C. The wall has a thermal conductivity of 0.9 W/m K.

6.2.2 Analogy with electrical conduction

Heat conduction can be compared with electrical conduction and equation [3] compared with Ohm's law for the current I through a resistor of resistance R when there is a potential difference V across it, i.e.

$$I = \frac{V}{R} \tag{4}$$

For heat transfer by conduction we can write equation [2] as:

$$Q = \frac{T_2 - T_1}{(x/kA)} = \frac{T_2 - T_1}{R} \tag{5}$$

The heat transfer rate is equivalent to electrical current I and the temperature difference to the electrical potential difference V and the thermal resistance for the plane layer is thus:

$$\text{thermal resistance } R = \frac{x}{kA} \tag{6}$$

Thermal resistance thus compares with electrical resistance expressed in terms of electrical conductivity σ as:

$$R = \frac{L}{\sigma A} \qquad [7]$$

or electrical resistivity ρ:

$$R = \frac{L}{(1/\rho)A} \qquad [8]$$

Thus thermal conductivity is equivalent to electrical conductivity or the reciprocal of electrical resistivity. The equivalence can thus be summarised as:

> Heat transfer rate is equivalent to electrical current, temperature difference is equivalent to potential difference and thermal conductivity is equivalent to electrical conductivity.

Example

Calculate the thermal resistance of a brick wall of area 15 m² and thickness 100 mm if the brick has a thermal conductivity of 1.2 W/m K.

Using equation [6]:

$$\text{thermal resistance} = \frac{x}{kA} = \frac{0.100}{1.2 \times 15} = 5.6 \times 10^{-3} \text{ K/W}$$

Revision

3   Calculate the thermal resistance per square surface metre of a furnace wall which is 220 mm thick and built of brick of thermal conductivity 0.87 W/m K.

6.2.3 Conduction through composite materials

In many situations involving heat conduction we have the heat being conducted through a series of layers, e.g. a furnace wall consisting of a layer of firebrick and a layer of insulating brick or the partition wall of a house consisting of two layers of plasterboard separated by a layer of glass-fibre insulation. Consider a plane wall which consists of three layers (Figure 6.2), each layer having a different thermal conductivity and thickness. We will assume that all the heat transfer takes place through the wall at right angles to the wall. Thus we have the same rate of energy input Q to each layer. For the first layer, using equation [3], we have:

$$Q = -k_1 A \frac{T_2 - T_1}{x_1} \qquad [9]$$

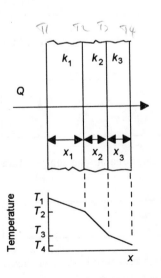

Figure 6.2   *Composite wall*

and for the second layer:

$$Q = -k_2 A \frac{T_3 - T_2}{x_2}$$

[10]

and for the third layer:

$$Q = -k_3 A \frac{T_4 - T_3}{x_3}$$

[11]

Rearranging these equations gives:

$$T_2 - T_1 = -\frac{Q}{A}\frac{x_1}{k_1}$$

[12]

$$T_3 - T_2 = -\frac{Q}{A}\frac{x_2}{k_2}$$

[13]

$$T_4 - T_3 = -\frac{Q}{A}\frac{x_3}{k_3}$$

[14]

Adding equations [12], [13] and [14] gives:

$$T_4 - T_1 = -\frac{Q}{A}\left[\frac{x_1}{k_1} + \frac{x_2}{k_2} + \frac{x_3}{k_3}\right]$$

[15]

$T_4 - T_1$ is the overall temperature difference across the composite wall; we thus do not need to know the temperatures of the intermediate surfaces in order to calculate the heat transferred. Equation [15] can be written as:

$$Q = -\frac{A}{\frac{x_1}{k_1} + \frac{x_2}{k_2} + \frac{x_3}{k_3}}(T_4 - T_1)$$

[16]

The term *overall heat transfer coefficient U* or, as it is often called, the *U-value* is defined as:

$$U = \frac{1}{\frac{x_1}{k_1} + \frac{x_2}{k_2} + \frac{x_3}{k_3}}$$

[17]

and relates the rate of heat transfer per unit area to the temperature difference. Thus equation [16] can be written as:

$$Q = -UA(T_4 - T_1)$$

[18]

Table 6.2 shows some typical U-values for composite plane layers used in buildings.

Table 6.2 U-values

Element	Composition	U-value, W/m^2 K
Solid wall	Brickwork 215 mm, plaster 15 mm	2.3
Cavity wall	Brickwork 102.5 mm, cavity 50 mm, brickwork 102.5 mm	1.6
Cavity wall	Brickwork 102.5 mm, cavity 25 mm, polystyrene board, 25 mm, aerated concrete block 100 mm, plaster 13 mm	0.58
Window	Single glazing	5.7
Window	Double glazing, airspace 20 mm	2.8

Example

The wall of a furnace consists of a layer of firebrick of thickness 200 mm with an outer layer of insulation of thickness 10 mm. The firebrick has a thermal conductivity of 0.7 W/m K and the insulator a thermal conductivity of 0.1 W/m K. Determine the rate of heat transfer per square metre of wall when the internal surface temperature is 600°C and the external surface temperature is 50°C.

Using equation [16]:

$$Q = -\frac{A}{\frac{x_1}{k_1} + \frac{x_2}{k_2}}(T_3 - T_1) = -\frac{1}{\frac{0.200}{0.7} + \frac{0.010}{0.1}}(50 - 600)$$

Hence $Q = 1426$ W.

Example

Calculate the heat loss per square metre through a wall with a U-value of 1.6 W/m^2 K when the internal surface temperature of the wall is 20°C and the external surface temperature is 0°.

Using equation [18]:

$$Q = -UA(T_4 - T_1) = -1.6 \times 1 \times (0 - 20) = 32 \text{ W}$$

Revision

4 A double-glazed window consists of two panes of glass of thickness 5 mm separated by an air space of thickness 20 mm. If the thermal conductivity of glass is 0.9 W/m K and that of air 0.03 W/m K, determine the heat loss per square metre of glazing when the internal temperature is 20°C and the external temperature is 0°C.

5   A double-glazed window has a U-value of 2.8 W/m² K. Determine the heat loss per square metre of glazing when the internal temperature is 20° and the external temperature is 0°C.

6.2.4 Composite layers: electrical analogue

In terms of the electrical analogue discussed in Section 6.2.2, when we have a plane composite wall in which the heat is conducted through first one layer and then another the electrical analogue is resistors in series (Figure 6.3). With series resistors we have the same current through each; analogous to the same rate of heat transfer through each of the layers. Applying Ohm's law to the circuit in Figure 6.3 gives:

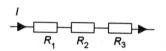

$$I = \frac{1}{R_1 + R_2 + R_3}V \qquad [19]$$

Figure 6.3   *Electrical analogue*

where V is the overall potential difference; analogous to the overall temperature difference for the heat transfer. We can compare this equation with equation [16] and the three thermal resistances of x_1/k_1A, x_2/k_2A and x_3/k_3A.

> For thermal resistances in series, the total resistance is equal to the sum of the resistances.

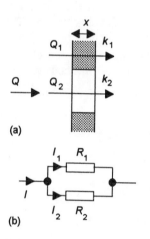

(a)

(b)

Figure 6.4   *Elements in parallel: (a) thermal, (b) electrical*

If we consider a wall containing a window, then heat is transferred at different rates through the wall and the window. The rate at which heat is transferred through the structure is the sum of the heat transfer rates through the wall and the window. This is comparable to an electrical network containing resistors in parallel (Figure 6.4); the total current is equal to the sum of the currents through the parallel branches, $I = I_1 + I_2$, and thus, applying Ohm's law gives:

$$I = \frac{V}{R_1} + \frac{V}{R_2}$$

For the thermal system described by Figure 6.4(a):

$$Q = Q_1 + Q_2$$

Applying equation [3] to each element, with element 1 having thermal conductivity k_1 and an area A_1 and element 2 with thermal conductivity k_2 and area A_2, gives:

$$Q = -k_1A_1\frac{T_1 - T_2}{x} - k_2A_2\frac{T_1 - T_2}{x}$$

$$= \frac{T_2 - T_1}{\frac{x}{k_1A_1}} + \frac{T_2 - T_1}{\frac{x}{k_2A_2}} \qquad [20]$$

With the thermal resistances being $x/k_1 A_1$ and $x/k_2 A_2$, the equation is comparable with equation [20].

> For parallel thermal resistors, the reciprocal of the total resistance is equal to the sum of the reciprocals of the individual resistances.

It is often simpler to consider heat transfer problems involving composites in terms of their electrical equivalents.

Example

The wall of a furnace consists of a layer of firebrick of thickness 200 mm with an outer layer of insulation of thickness 10 mm. The firebrick has a thermal conductivity of 0.7 W/m K and the insulator a thermal conductivity of 0.1 W/m K. Determine the rate of heat transfer per square metre of wall when the internal surface temperature is 600°C and the external surface temperature is 50°C.

This is a repeat of an earlier example from this chapter but with the solution considered in 'electrical' terms. The arrangement is of two thermal resistances in series. The firebrick has a thermal resistance of $x_1/k_1 A = 0.200/0.7 = 0.286$ K/W and the insulation a thermal resistance of $x_2/k_2 A = 0.010/0.1 = 0.10$ K/W. Thus the total series resistance is $0.286 + 0.10 = 0.386$ K/W. Using the thermal equivalent of Ohm's law:

$$Q = \frac{\text{temp. difference}}{\text{resistance}} = \frac{550}{0.386} = 1425 \text{ W}$$

Example

Draw the equivalent electrical circuit for the heat transfer through a plane wall consisting of one material riveted to another in the manner shown in Figure 6.5.

The arrangement consists of materials 1 and 2 in series, since the same heat flow occurs through both. This series arrangement is in parallel with the rivet material since the total heat flow is the sum of that transmitted through each element. The equivalent electrical circuit is thus as shown in Figure 6.6.

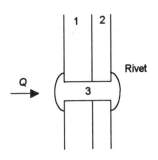

Figure 6.5   *Example*

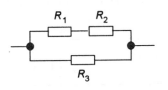

Figure 6.6   *Example*

Revision

6   State what the equivalent electrical circuits would be for heat transfer by conduction through (a) a single brick wall containing a window, (b) a cavity brick wall containing a window.

7   The door of a freezer has an interior layer of plastic 5 mm thick, an insulating foam of thickness 80 mm and an outer layer of steel 1 mm

thick. The surface area of the door is 1.6 m². Determine the heat transfer by conduction though the door when the temperature inside the refrigerator is –15°C and the exterior temperature is 18°C. The thermal conductivity of the plastic is 3.95 W/m K, of the insulating foam 0.05 W/m K and of the steel 50 W/m K.

6.2.5 Heat conduction through cylindrical layers

For a plane wall, the area perpendicular to the heat flow is the same throughout the entire heat flow path. This is not the case for a cylinder. Consider a cylinder of length L, internal radius r_1 and external radius r_2 (Figure 6.7). We will assume that there is negligible heat flow along the length of the cylinder or circumferentially round the cylinder wall and that all the heat flow takes place in a radial direction. Consider a thin cylindrical element of radius r and thickness δr. The area through which heat flows is $2\pi r L$. If the temperature difference across this layer is δT then, using equation [1], we can write for the rate of heat flow Q:

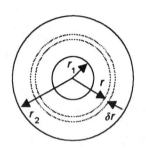

Figure 6.7   *Conduction through a cylindrical wall*

$$Q = -k(2\pi rL)\frac{\delta T}{\delta r}$$

where k is the thermal conductivity of the cylinder material. If we consider infinitesimally thin layers then we can write this equation as:

$$Q = -k(2\pi rL)\frac{\mathrm{d}T}{\mathrm{d}r}$$

We have a large number of such layers in series between radius r_1 and radius r_2 and thus to determine the heat transfer from inside the cylinder to outside we take the sum of resistances of each of the elements, i.e. we integrate the expression. Thus, if T_1 is the temperature of the inside wall of the cylinder and T_2 that of the outside wall:

$$Q\int_{r_1}^{r_2} \frac{1}{r}\,\mathrm{d}r = -2\pi kL \int_{T_1}^{T_2} \mathrm{d}T$$

$$Q(\ln r_2 - \ln r_1) = -2\pi kL(T_2 - T_1)$$

and thus:

$$Q = 2\pi kL \frac{T_2 - T_1}{\ln(r_2/r_1)} \qquad [22]$$

Comparing this equation with the electrical analogue of a resistor, then the thermal resistance of the cylinder wall is:

$$\text{thermal resistance} = \frac{\ln(r_2/r_1)}{2\pi kL} \qquad [23]$$

If we have radial conduction through a multi-layer cylindrical wall, then we can consider it to be represented by a series of resistors in series and so the total resistance is the sum of the resistances of each layer.

Example

A copper pipe carries hot water at 70°C and has an external radius of 75 mm. It is lagged to an overall radius of 125 mm. The surface temperature of the lagging is found to be 25°C. Assuming that the inner surface of the lagging is at the temperature of the hot water, determine the rate of heat loss per metre length of pipe. The thermal conductivity of the lagging is 0.09 W/m K.

Using equation [22]:

$$Q = 2\pi \times 0.09 \times 1 \frac{70 - 25}{\ln(0.125/0.075)} = 49.8$$

Example

A pipe has an internal diameter of 50 mm and an external diameter of 100 mm, the pipe material having a thermal conductivity of 50 W/m K. The pipe is lagged by a cylindrical jacket of thickness 15 mm, the lagging having a thermal conductivity of 0.04 W/m K. If the pipe carries steam which gives its internal wall a temperature of 150° and the outer surface of the lagging is at 20°, what will be the rate of heat loss per metre length of pipe?

The thermal resistance of the pipe is given by equation [23] as:

$$\text{thermal resistance} = \frac{\ln(r_2/r_1)}{2\pi k L} = \frac{\ln(0.050/0.025)}{2\pi \times 50 \times 1}$$

$$= 2.21 \times 10^{-3} \text{ K/W}$$

The thermal resistance of the lagging is:

$$\text{thermal resistance} = \frac{\ln(r_2/r_1)}{2\pi k L} = \frac{\ln(0.065/0.050)}{2\pi \times 0.04 \times 1}$$

$$= 1.044 \text{ K/W}$$

Thus the total thermal resistance is 1.046 K/W and so:

$$Q = \frac{\text{overall temp. difference}}{\text{total resistance}} = \frac{150 - 20}{1.046} = 124.3 \text{ W}$$

Revision

6   A steam pipe has a surface temperature of 200°C and a diameter of 150 mm. It is lagged to an overall diameter of 250 mm, the lagging having a thermal conductivity of 0.09 W/m K. If the surface

temperature of the lagging is 50°C, determine the rate of heat loss from the pipe per metre length.

7   A pipe has an internal diameter of 30 mm and a wall thickness of 3 mm. It is lagged with a cylindrical jacket of thickness 25 mm. The thermal conductivity of the pipe material is 50 W/m K and that of the insulation 0.03 W/m K. Determine the total thermal resistance per metre length of pipe to heat conduction.

6.3 Convection

Because most fluids have low thermal conductivities, the heat transferred through fluids by direct conduction is relatively small. Instead we have to consider heat transfer by convection. For example, for a hot object suspended in still air, a thin layer of air surrounding the object is heated by conduction. As a result of this air increasing in temperature, it expands and so its density decreases. The hot air thus rises and is replaced by cool air. This movement of the fluid is termed *convection* and is the mechanism by which heat is transferred from the hot object to the surrounding air. The term *natural convection* is used when the cooling process arises purely from such density changes. However, if the motion of the fluid is enhanced by means of a fan or pump, the term *forced convection* is used.

6.3.1 Heat transfer coefficient

Convection is concerned with the movement of fluids and there are many factors which determine the rate at which fluids move (see Chapter 7) and hence the rate at which convection transfers heat. However, a general principle which can be used to describe convective heat transfer was proposed by Newton as: the rate at which heat is transferred by convection is proportional to the temperature difference between a body and the surrounding fluid. This is expressed as:

$$\text{rate of convective heat transfer } Q = hA(T_1 - T_2) \qquad [24]$$

where A is the surface area of the body at temperature T_1, T_2 is the temperature of the surrounding fluid and h is a constant called the *surface heat transfer coefficient*. The value of h depends on the fluid concerned, whether the convective fluid flow is free or forced and whether it is laminar or turbulent. For natural convection h is typically between about 4 and 50 W/m² K, for forced convection in air between about 10 and 550 W/m² K and for forced convection in liquids between about 100 and 5500 W/m² K.

In terms of the electrical analogue, the thermal resistance for convective heat transfer from an area is:

$$\text{thermal resistance} = \frac{1}{hA} \qquad [25]$$

Example

A beaker of hot water is in a room where the surroundings are at 20°C. How will its rates of cooling by convection at 60°C and 30°C compare?

As equation [24] indicates, the rate of cooling is proportional to the temperature difference. Thus the ratio of the rates of cooling at the two temperatures will be:

$$\frac{\text{rate of coooling at } 60°C}{\text{rate of cooling at } 30°C} = \frac{60-20}{30-20} = 4$$

Example

A flat plate has a surface area of 0.1 m². What is the convective thermal resistance of the surface if the surface heat transfer coefficient is 10 W/m² K?

Using equation [25]:

$$\text{thermal resistance} = \frac{1}{hA} = \frac{1}{10 \times 0.1} = 1 \text{ K/W}$$

Revision

10 What is the convective thermal resistance per metre length of the outside surface of a pipe with an external diameter of 150 mm if the surface heat transfer coefficient is 15 W/m² K?

11 Hot water is carried by two pipes, one of diameter 100 mm and the other of diameter 200 mm. If the surface temperatures and surface heat transfer coefficients are the same for both pipes and they are in surroundings at the same temperature, how does the rate of convective transfer of heat from the two pipes compare?

6.3.2 Fluids separated by a solid boundary

So far in this chapter, heat transfer by conduction and convection have been considered separately. However, in many situations both often occur together. Thus in considering the heat loss through a house wall we have a situation where there is a fluid, air, on both sides of the wall and we have convective heat transfer from the warm air in the house to the house wall, heat transfer by conduction through the wall and then convective heat transfer from the outside of the wall to the surrounding air. We thus have heat transfer between fluids separated by a solid wall. The situation is thus one of three series thermal resistances, the convective resistance of the fluid/solid interface plus the conductive resistance of the solid plus the convective resistance of the solid/fluid interface.

For fluids separated by a plane solid boundary, as in Figure 6.8, the total thermal resistance is:

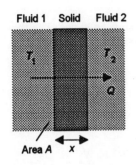

Figure 6.8   *Fluids separated by a plane solid boundary*

total resistance = convective resistance from fluid 1 to solid + conductive resistance of the solid + convective resistance from solid to fluid 2

If h_1 is the surface heat transfer coefficient for fluid 1, h_2 the surface heat transfer coefficient for fluid 2 and k the thermal conductivity of the solid wall, then:

$$\text{total resistance } R = \frac{1}{h_1 A} + \frac{x}{kA} + \frac{1}{h_2 A} \qquad [26]$$

Thus the rate of transfer of heat from fluid 1 at temperature T_1 to fluid 2 at temperature T_2 is:

$$Q = \frac{T_1 - T_2}{R} \qquad [27]$$

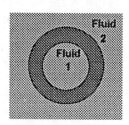

For a cylindrical boundary between two fluids (Figure 6.9), with the inner surface of the solid having a radius r_1, the external surface a radius r_2 and the boundary a length L, the inner surface area is $2\pi r_1 L$ and the external surface area is $2\pi r_2 L$. Thus the total resistance is:

Figure 6.9   *Fluids separated by a cylindrical boundary*

$$\text{total resistance} = \frac{1}{2\pi r_1 L h_1} + \frac{\ln(r_2/r_1)}{2\pi L k} + \frac{1}{2\pi r_2 L h_2} \qquad [28]$$

Example

An insulated pipe carries steam at 300°C and has an internal diameter of 35 mm, a wall thickness of 3 mm and an insulation thickness of 25 mm. If heat transfer along the pipe is negligible, determine the radial rate of heat transfer per metre length of pipe when the external air temperature is 20°. The thermal conductivity of the pipe material is 50 W/m K, the thermal conductivity of the insulation is 0.03 W/m K, the inside surface heat transfer coefficient is 10 000 W/m² K and the external surface heat transfer coefficient is 10 W/m² K.

The situation is one of three thermal resistors in series. The thermal resistance for conduction through the pipe wall is:

$$\text{resistance} = \frac{\ln(0.0205/0.0175)}{2\pi \times 1 \times 50} = 5.0 \times 10^{-4} \text{ K/W}$$

The thermal resistance for conduction through the insulation is:

$$\text{resistance} = \frac{\ln(0.0455/0.0205)}{2\pi \times 1 \times 0.03} = 4.2297 \text{ K/W}$$

The thermal resistance for the internal surface is:

$$\text{resistance} = \frac{1}{2\pi \times 0.0175 \times 1 \times 10\,000} = 9.1 \times 10^{-4} \text{ K/W}$$

The thermal resistance for the external surface is:

$$\text{resistance} = \frac{1}{2\pi \times 0.0455 \times 1 \times 10} = 0.3498 \text{ K/W}$$

The total resistance is thus 4.5809 K/W and hence, for a temperature difference of $(300 - 20) = 280°C$ the heat transfer is 61.1 W.

Revision

12 A cavity wall consists of two 105 mm thick brick walls separated by a cavity which is filled with a 25 mm thick sheet of expanded polystyrene. The inner wall has a 13 mm thickness of plaster. Determine the heat transfer per square metre of wall when the internal air temperature is 21°C and the external temperature 0°C. The thermal conductivity of the brick used for the outer wall is 0.8 W/m K, the thermal conductivity of the brick used for the inner wall is 0.6 W/m K, the thermal conductivity of the expanded polystyrene is 0.04 W/m K, the thermal conductivity of the plaster is 0.2 W/m K, the internal surface heat transfer coefficient is 8 W/m² K and the external surface heat transfer coefficient is 17 W/m² K.

6.4 Radiation

Unlike conduction and convection, heat transfer by radiation does not require a medium for its transmission, being able to travel through a vacuum. Heat transfer by radiation is the transfer of energy by means of electromagnetic waves. When radiation is incident on a body, some may be absorbed and result in an increase in temperature, some may be transmitted through the body and leave it unaltered, and some may be reflected and leave the body unaltered. The fraction of incident energy that is absorbed by a surface is called the *absorptivity a*, the fraction transmitted the *transmissivity t* and the fraction reflected the *reflectivity r*. A body for which the absorptivity is 1 or 100%, i.e. all the radiation is absorbed and none is reflected or transmitted, is termed a *black body*. A black body absorbs all the radiation incident on it, regardless of the wavelength of the radiation. A black body at some temperature must be also be an emitter of radiation, if it kept on just absorbing radiation its temperature would go on rising indefinitely. When it is at temperature T in thermal equilibrium with its surroundings then the rate at which it absorbs energy must be equal to the rate at which it emits energy. A black body emits radiation over the entire range of wavelengths.

The rate at which energy is absorbed or emitted by a black body is proportional to the fourth power of its absolute temperature T. This is known as the *Stefan-Boltzmann law* and can be expressed as:

rate of energy absorbed or emitted/unit area $q = \sigma T^4$ [29]

with σ being called the Stefan-Boltzmann constant and equal to 5.67×10^{-8} W/m^2 K.

For a black body of surface area A and at a temperature T in surroundings at a temperature T_s, the rate at which the body emits radiation is $A\sigma T^4$. If we consider it to be completely enclosed by the black body of area A_s representing the surroundings (Figure 6.10) then the rate at which the surroundings emit energy is $A_s\sigma T_s^4$. But not all this energy will fall on the black body, only the fraction A/A_s falls on the black body with the rest being reabsorbed by the surroundings. Thus the black body receives energy from the surroundings of $A\sigma T_s^4$. Thus the net rate of transfer of energy to the surroundings is:

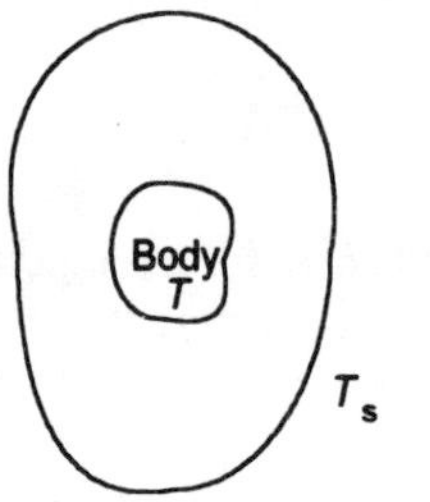

Figure 6.10   *Black body in a black enclosure*

$$\text{net rate of transfer of energy} = A\sigma(T^4 - T_s^4) \qquad [30]$$

In problems concerning the transfer of heat radiation between two surfaces, we need to take into account the fraction of the radiation emitted by each surface which will impinge on the other.

For a black body, equation [29] gives the rate of energy absorption; for a non-black body not all the energy impinging on it will be absorbed and thus to obtain the energy absorbed we have to scale down the energy absorbed by multiplying the energy that a black body would absorb by the absorptivity, i.e.

$$q = a\sigma T^4 \qquad [31]$$

For a black body, equation [29] gives the rate of energy emission; to obtain the energy that would be emitted by a non-black body the energy emitted by a black body needs scaling by multiplying by the *emissivity ε*, i.e.

$$q = \varepsilon\sigma T^4 \qquad [32]$$

The *emissivity* of a body is defined as the radiation emitted by the body as a fraction of that which would have been emitted by a black body at the same temperature.

Table 6.3 gives typical emissivity and absorptivity values for materials at about 20°C. The term *grey body* is used for a non-black body which has the same emissivity and absorptivity at every wavelength. Many real surfaces do approximate to grey bodies.

A *black* body is a perfect absorber or emitter of radiant heat energy at all wavelengths. A *grey* body is a non-black body which has the same absorptivity or emissivity at all wavelengths.

Table 6.3 Emissivity and absorptivity values

Surface	Emissivity	Absorptivity
Aluminium, highly polished	0.04	0.20
Asphalt	0.95	0.90
Concrete	0.90	0.65
Building brick (red)	0.90	0.60
White gloss painted surface	0.90	0.30
Black-matt painted surface	0.90	0.90

Example

A 1 kW electric fire element has an emitting surface which is 250 mm long and of diameter 18 mm. If you can assume that the element behaves as a black body and that 95% of the electrical energy supplied is dissipated as radiation, calculate the temperature of the element if the surroundings are at 20°C.

Using equation [30], i.e. rate of loss of energy $= A\sigma(T^4 - T_s^4)$, then:

$$0.95 \times 1000 = 0.250 \times \pi \times 0.018 \times 5.67 \times 10^{-8}(T^4 - 293^4)$$

Hence $T = 1045K$ or 772°C.

Example

The sun has a radius of 6.98×10^8 m and behaves as a black body at a temperature of 5800 K. The mean distance of the earth from the sun is 1.50×10^{11} m. Determine the rate at which energy is received per square metre of the earth's surface if the effects of the earth's atmosphere are neglected.

Using equation [29], the rate at which energy is emitted by the sun is:

$$Q = A\sigma T^4 = 4\pi \, (6.98 \times 10^8)^2 \times 5.67 \times 10^{-8} \times 5800^4$$

$$= 3.93 \times 10^{26} \text{ W}$$

This energy spreads out, at the earth's distance, over a sphere of surface area $4\pi(1.50 \times 10^{11})^2$. Hence the rate at which energy is received by 1 m² is:

$$\text{rate at which energy is received} = \frac{3.93 \times 10^{26}}{4\pi(1.50 \times 10^{11})^2}$$

$$= 1.39 \times 10^3 \text{ W}$$

Example

A solar panel has a heat collecting surface with an area of 2 m², an absorptivity of 0.92 and an emissivity of 0.15. The convective surface heat transfer coefficient is 3 W/m² K. The rate at which energy is received from the sun is 800 W/m². Determine the net rate of energy collection when the installation is at a temperature of 60°C and the surroundings are at 18°C.

The rate at which energy is absorbed by the panel is $0.92 \times 800 = 736$ W/m². The radiation loss per square metre is $\varepsilon \sigma T^4 = 0.15 \times 5.67 \times 10^{-8} \times 333^4 = 104.6$ W/m². The convection loss per square metre is $3 \times (60 - 18) = 126$ W/m². Thus the net rate of energy collection per square metre is $736 - 104.6 - 126 = 505.4$ W/m² and thus for the area of 2 m² the energy collected is 1.01 kW.

Revision

13   A black body has a surface area of 0.2 m² and is at 540°C. Determine the rate of energy emission from the body.

14   Determine the rate of energy emission per square metre of a body at 50°C if it has an emissivity of 0.3.

15   A sphere of diameter 30 mm is suspended in a vacuum with negligible heat conduction occurring along its suspension. At what rate must energy be supplied to the sphere to maintain it at a temperature of 500 K if the surroundings are at 300K?

16   A pipe with an external diameter of 50 mm carries hot water. The external surface of the pipe is at a temperature of 80°C and the surrounding air is at 20°C. If the emissivity of the pipe surface is 0.3 and the surface heat transfer coefficient is 3 W/m² K determine the heat loss per metre length of pipe.

Problems

1   Calculate the rate of heat input required to maintain a temperature difference of 2°C between the two sides of a metal sheet of area 2 m² and thickness 10 mm if the sheet has a thermal conductivity of 50 W/ m K.

2   Determine the rate at which heat is transferred along a lagged copper rod of length 500 mm and cross-sectional area 1000 mm² if the temperature difference between the two ends is 100°C. Take the thermal conductivity of copper as 380 W/m K.

3   The cavity wall of a house consists of brick of thickness 120 mm, a 50 mm air cavity and then another brick section of thickness 120 mm. What is the rate at which heat is conducted through a cavity wall of area 15 m² when the difference in temperature

between the inner surface and outer surfaces of the wall is 20°C. Take the thermal conductivity of the brick to be 1.0 W/m K and the air 0.03 W/m K.

4   A brick wall of thickness 250 mm is coated with a layer of concrete of thickness 50 mm. If the temperature difference between the two sides of the wall is 25°C, determine the rate of transmission of heat by conduction through one square metre of wall. Take the thermal conductivity of the brick to be 0.7 W/m K and that of the concrete to be 0.9 W/m K.

5   Determine the rate of heat transfer through a wall of area 100 m² when there is a temperature difference of 20°C between the inner and outer surfaces of the wall and the wall has a U-value of 1.5 W/m² K.

6   A plain cavity wall has a U-value of 0.91 W/m² K. What thickness of expanded polystyrene should be introduced into the cavity to reduce the U-value to 0.6 W/m² K. The polystyrene has a thermal conductivity of 0.033 W/m K. Hint: consider the thermal resistance that has to be introduced into the 'series circuit' for 1 m² of wall.

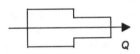

Figure 6.11   *Problem 7*

7   State what the equivalent electrical circuits would be for heat transfer by conduction through (a) a lagged metal rod where there is a change in cross-section (Figure 6.11), (b) a composite panel consisting of two sheets of material riveted together.

8   Two lagged metal bars A and B are connected (a) in series, (b) in parallel. With the bars in series, the free end of A is at 80°C and the free end of B at 20°C. When the bars are in parallel, the ends of each bar are at 80°C and 20°C. Determine the ratio of the total rate of heat transfer by conduction in the series arrangement to that in the parallel. The thermal conductivity of A is 400 W/m K and that of B is 200 W/m K.

9   For a domestic hot water system, a copper pipe carries hot water at 70°C and has an external diameter of 150 mm and is lagged to an overall diameter of 500 mm. If the surface temperature of the lagging is 20°C determine the rate of heat loss per metre length of pipe if it can be assumed that the inner surface of the lagging is at the hot water temperature. The thermal conductivity of the lagging is 0.09 W/m K.

10   A lagged pipe has an external diameter of 200 mm. What is the convective thermal resistance per metre length of the outside lagging surface if the surface heat transfer coefficient is 10 W/m² K?

11   To keep an electrical component cool during use, a fan may be used to blow air over it and additionally metal baffle plates may be attached to it. Explain the reasons behind such methods.

12 An insulated pipe carries steam at 200°C and has an internal diameter of 50 mm, a wall thickness of 25 mm and an insulation thickness of 13 mm. If heat transfer along the pipe is negligible, determine the radial rate of heat transfer per metre length of pipe when the external air temperature is 20°. The thermal conductivity of the pipe material is 50 W/m K, the thermal conductivity of the insulation is 0.04 W/m K, the inside surface heat transfer coefficient is 15 W/m² K and the external surface heat transfer coefficient is 10 W/m² K.

13 A cavity wall consists of two 100 mm thick brick walls separated by a cavity. The inner wall has a 15 mm thickness of plaster. Determine the heat transfer per square metre of wall when the internal air temperature is 20°C and the external temperature 5°C. The thermal conductivity of the brick is 0.7 W/m K, the thermal conductivity of the plaster is 0.4 W/m K, the cavity has a U-value of 6 W/m² K, the internal surface heat transfer coefficient is 10 W/m² K and the external surface heat transfer coefficient is 20 W/m² K.

14 A 1.5 kW radiant heater consists of a heating element which is a silica cylinder of length 600 mm and diameter 10 mm. assuming that the heater behaves as a black body and that there is 100% transformation of electrical energy to heat radiation, determine the temperature of the heater. You can neglect any consideration of the input of radiation to the heater from the surroundings.

15 Determine the net rate of energy absorption by a solar panel if it has an area of 1.5 m², an absorptivity of 0.96, an emissivity of 0.20, a convective surface heat transfer coefficient of 3 W/m² K, has a surface temperature of 67°C in surroundings at a temperature of 20°C, and the rate of energy being received from the sun is 500 W/m².

16 A piece of tungsten wire of length 12 mm and diameter 0.7 mm is at a temperature of 2000°C. The emissivity of the tungsten is 0.35. Determine the net heat energy radiated per second by the wire when the surroundings are at a temperature of 20°C.

7 Fluid flow

7.1 Introduction

This chapter is about the energy transfer that occurs as a result of the flow of 'real' fluids. An 'ideal' fluid flowing through a pipe will have the entire fluid across a cross-section flowing with the same velocity since the concept of an ideal fluid assumes that there are no viscous friction forces opposing the motion. In practice, real fluids have viscosity and forces are needed to overcome viscous forces and move a fluid through a pipe. With an ideal fluid, when a shaft rotates in a lubricated journal bearing (Figure 7.1), the lubricant fluid would put up no opposition to the motion. There is a tangential force acting on the lubricant at the surface of the shaft and thus the lubricant is acted on by shear forces. An ideal fluid does not resist deformation from the shear forces. However, a real fluid resists the deformation resulting from shear forces and thus there is opposition to the rotation of the shaft. The property of a fluid which is responsible for its resistance to motion and deformation is called its *viscosity*.

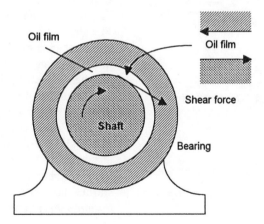

Figure 7.1 *Shaft rotating in a lubricated bearing*

7.2 Viscosity

A fluid consists of a large number of molecules in random motion. When a fluid is made to flow smoothly over a stationary surface, the random velocity of the molecules has superimposed on it an orderly velocity in a particular direction. The molecular layer in contact with the surface is at rest, being held to it by intermolecular forces. Subsequent layers move with increasing velocities as the distance from the stationary surface is

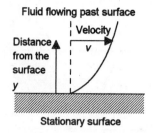

Fluid flowing past surface

Figure 7.2   *Velocity gradient*

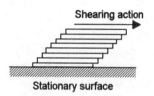

Figure 7.3   *Shearing a pack of cards*

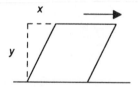

Figure 7.4   *Shear strain*

increased. There is thus a velocity gradient at right angles to the surface (Figure 7.2). There is a shearing action between adjacent layers. You can think of the situation being rather like a deck of cards face down on a table. The top card is acted on by a shearing force and slides over the one below it, dragging it along to a lesser extent, and so on down the pack with the last card on the table remaining at rest (Figure 7.3). The reason all the cards do not move with the top card is that the stationary surface exerts a frictional force opposing the motion of the card in contact with it and this card exerts a frictional force on the card immediately above it and so on for subsequent cards. With the molecules in a fluid, the random velocity element of their velocity means that the faster moving layer will lose some of its molecules and gain some of the molecules from a slower moving layer. This has the effect of lowering the average velocity of the layer. Since a velocity change means a force we have a drag force acting on the layer. This drag is what is termed *viscosity*.

For a fluid, the velocity gradient (dv/dy) is proportional to the shearing force per unit area (F/A), i.e. the shear stress, responsible for sliding one layer of liquid over another. This relationship is expressed as:

$$\frac{F}{A} = \eta \frac{dv}{dy} \qquad [1]$$

where η is called the *coefficient of viscosity* or the *dynamic viscosity*. The units of dynamic viscosity are $(N/m^2)(m)/(m/s) = N\ s/m^2$ or Pa s. The shear strain for a sheared material (Figure 7.4) is x/y. Since v is the rate of change of x with time then the velocity gradient dv/dy is the time rate of change of shear strain. Thus for a fluid, equation [1] states that the shear stress is proportional to the time rate of change of shear strain. This is different to the relationship for a solid where the shear stress is proportional to the shear strain, not its rate of change with time.

Fluids for which the shear stress is directly proportional to the time rate of shear strain are called *Newtonian fluids*, i.e. the dynamic viscosity is a constant. Water and oils are examples of Newtonian fluids. For some fluids the shear stress may not be directly proportional to the time rate of strain; these are called *non-Newtonian fluids*. Non-drip paint is an example of a non-Newtonian fluid. This has the property that the viscosity decreases as the time rate of shear strain is increased. Thus when the paint is brushed on and high strain rates occur, it flows more easily then when left to flow on the surface under its own weight and so lower strain rate.

Many equations in fluid mechanics include the combination of η/ρ, where ρ is the density of the fluid. This combination is termed the *kinematic viscosity*:

$$\text{kinematic viscosity} = \frac{\eta}{\rho} \qquad [2]$$

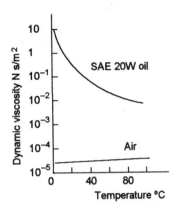

Figure 7.5 *Effect of changes in temperature on dynamic viscosities*

The units of kinematic viscosity are $(N\ s/m^2)/(kg/m^3)$. We can write, by using $F = ma$, the kg unit as $N/(m/s^2)$ and thus the units of kinematic viscosity are $(N\ s/m^2)/(N\ s^2/m)/(m^3) = m^2/s$.

Table 7.1 gives values of the dynamic and kinematic viscosities of some fluids at 20°C and under normal atmospheric pressure. Temperature has a considerable effect on viscosity. However, the viscosity of gases and most liquids increases only a little with an increase in pressure. Figure 7.5 illustrates how the viscosity changes with temperature. The viscosity of liquids decreases with temperature in a roughly exponential manner:

$$\eta \approx a\ e^{-bT} \qquad\qquad [3]$$

where T is the absolute temperature and a and b are constants. The viscosity of gases increases with temperature; an approximate relationship that is often used is:

$$\frac{\eta}{\eta_0} = \left(\frac{T}{T_0}\right)^n \qquad\qquad [4]$$

where η is the dynamic viscosity at temperature T, η_0 the dynamic viscosity at temperature T_0 and n a constant. For air, n is about 0.7.

Table 7.1 Values of dynamic and kinematic viscosities at 20°C and normal atmospheric pressure

Fluid	Dynamic viscosity N s/m²	Kinematic viscosity m²/s
Air	1.8×10^{-5}	1.1×10^{-4}
Petrol	2.9×10^{-4}	4.2×10^{-7}
Water	1.0×10^{-3}	1.0×10^{-6}
SAE 20 oil	1.2×10^{-1}	2.0×10^{-4}
Glycerine	1.5	1.2×10^{-3}

7.2.1 Flow between plates

Consider the flow induced in a fluid between a fixed lower plate and an upper plate moving steadily with a velocity v (Figure 7.6). The fluid layers in contact with each of the plates have the same velocities as the plates and so a velocity gradient will be produced in the fluid. Because the upper plate has a steady velocity, the velocity is proportional to the distance from the fixed plates and so the velocity gradient is constant. Thus $dv/dy = v/h$, where h is the separation between the plates, and the force F needed to maintain the upper plate in motion with a constant velocity v, i.e. the force needed to overcome the viscous drag, is given by equation [1] as:

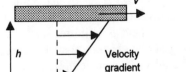

Figure 7.6 *Flow between plates*

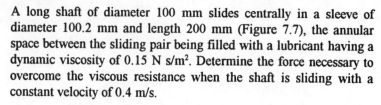

$$\frac{F}{A} = \eta \frac{v}{h} \qquad\qquad [4]$$

where A is the plate area and η the dynamic viscosity of the fluid.

Example

A long shaft of diameter 100 mm slides centrally in a sleeve of diameter 100.2 mm and length 200 mm (Figure 7.7), the annular space between the sliding pair being filled with a lubricant having a dynamic viscosity of 0.15 N s/m². Determine the force necessary to overcome the viscous resistance when the shaft is sliding with a constant velocity of 0.4 m/s.

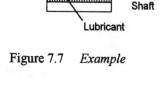

Figure 7.7   *Example*

The velocity gradient across the lubricant film is 0.4/0.0001 (m/s)/m and the area over which the shear force is applied is $\pi \times 0.100 \times 0.200$ m². Thus, using equation [4]:

$$\frac{F}{\pi \times 0.100 \times 0.200} = 0.15\frac{0.4}{0.000}$$

and so the force F necessary to overcome the viscous resistance is 37.7 N.

Revision

1   Two plates are separated by a 5 mm layer of lubricant. The lower plate is stationary and the upper plate moves with a constant velocity of 3 m/s. Determine the shear stress in the oil if it has a dynamic viscosity of 0.2 N s/m².

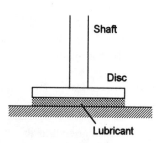

Figure 7.8   *Revision problem 2*

2   Figure 7.8 shows a disc being rotated by a shaft. The disc is positioned close to, and parallel to, a solid surface with the space between them being filled with a lubricant layer. When the disc is rotated at 3 rad/s, what is the ratio of the shear stress in the oil at a radial distance of 20 mm from the shaft axis to that at 30 mm?

7.3 Power loss with bearings

Consider *parallel bearings* with two flat surfaces (Figure 7.9) and between which there is a constant thickness layer of lubricant of thickness h. If one surface is moving relative to the other with a velocity v then, if we assume that the velocity gradient is uniform, we have a velocity gradient of v/h. If the bearing surface has an area A then the viscous resistive force F to the motion is given by equation [4] as:

$$F = \eta \frac{v}{h} A \qquad\qquad [5]$$

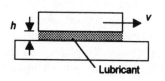

Figure 7.9   *Parallel bearing*

The distance moved per second by the point of application of the force is v and thus the work done per second, i.e. the power, is Fv. Thus the power loss as a consequence of the viscous resistance is:

$$\text{power} = Fv = \frac{\eta v^2 A}{h} \qquad [6]$$

Example

A long shaft of diameter 100 mm slides centrally in a sleeve of diameter 100.2 mm and length 200 mm (Figure 7.10), the annular space between the sliding pair being filled with a lubricant having a dynamic viscosity of 0.15 N s/m^2. Determine the power necessary to overcome the viscous resistance when the shaft is sliding with a constant velocity of 0.4 m/s.

This is an extension of the example in the previous section. The bearing area is $\pi \times 0.100 \times 0.200$ m^2. Thus, using equation [6]:

$$\text{power} = \frac{\eta v^2 A}{h} = \frac{0.15 \times 0.4^2 \times \pi \times 0.100 \times 0.200}{0.0001} = 15.1$$

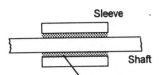

Sleeve

Shaft

Lubricant

Figure 7.10   *Example*

Revision

3   A shaft of diameter 60.0 mm is being pushed with a constant velocity of 0.5 m/s through a concentric bearing sleeve of diameter 60.2 mm and 300 mm long. The space between the shaft and the cylinder is filled with a lubricant of dynamic viscosity 0.2 N s/m^2. Determine the resistant force and the power required.

7.3.1 Power loss with collar and footstep bearings

Consider a collar thrust bearing of external radius r_2 and internal radius r_1 with the load transmitted to an annular surface in the bearing housing through a lubricant of thickness h (Figure 7.11). The shaft is rotated with an angular velocity ω. The tangential velocity acting on the lubricant film will, since $v = r\omega$, depend on the radius considered. For an annular element of radius r and thickness δr:

$$F = \eta \frac{dv}{dy} A = \eta \frac{r\omega}{h} 2\pi r \delta r \qquad [7]$$

Hence the torque acting on that element is:

$$\text{torque on element} = \eta \frac{r\omega}{h} 2\pi r \delta r \times r = \frac{2\pi \eta \omega}{h} r^3 \delta r \qquad [8]$$

The total torque T acting on the collar will be thus the sum of all the torques acting on each of elements between radii r_1 and r_2 and thus, when we consider elements of infinitesimally small thickness:

$$T = \frac{2\pi \eta \omega}{h} \int_{r_1}^{r_2} r^3 \, dr = \frac{\pi \eta \omega}{2h} (r_2^4 - r_1^4) \qquad [9]$$

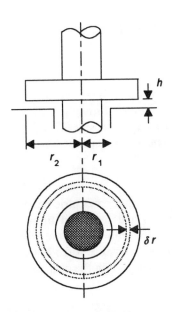

r_2 r_1

δr

Figure 7.11   *Collar bearing*

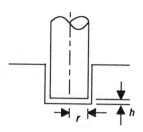

Figure 7.12   *Footstep bearing*

The work done when a torque T moves its point of application through an angle θ is $T\theta$ (equation [54], Chapter 4) and thus when there is an angular velocity ω then the rate of working, i.e. power, is $T\omega$. Thus the power necessary to overcome the viscous resistance is:

$$\text{power} = T\omega = \frac{\pi\eta\omega^2}{2h}(r_2^4 - r_1^4) \qquad [10]$$

For a *footstep thrust bearing* of radius r (Figure 7.12), equation [10] reduces to:

$$\text{power} = T\omega = \frac{\pi\eta\omega^2 r^4}{2h} \qquad [11]$$

Example

A collar thrust bearing is used with a shaft rotating at 30 rev/s and with the load transmitted to an annular surface of internal radius 40 mm and external radius 70 mm via a lubricant layer of thickness 0.4 mm and dynamic viscosity 0.1 N s/m². Determine the power required to overcome the viscous torque exerted by the lubricant.

Using equation [10]:

$$\text{power} = \frac{\pi\eta\omega^2}{2h}(r_2^4 - r_1^4)$$

$$= \frac{\pi \times 0.1 \times (2\pi \times 30)^2}{2 \times 0.0004}(0.070^4 - 0.040^4) = 299 \text{ W}$$

Revision

4   A vertical shaft of diameter 75 mm rotates at 1000 rev/min in a footstep bearing, the thrust being transmitted through a film of lubricant of thickness 0.2 mm and dynamic viscosity 0.2 N s/m². Determine the power required to overcome the viscous resistance.

7.3.2 Power loss with journal bearings

Consider a journal bearing, i.e. a shaft rotating in a concentric bearing, with the shaft rotating at angular velocity ω and separated from the bearing by a constant clearance c which is filled with a lubricant with a dynamic viscosity η (Figure 7.13). The viscous resistance force F acting tangentially at the circumference of the shaft, length L in the bearing, is:

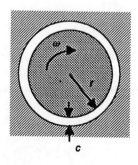

Figure 7.13   *Journal bearing*

$$F = \eta\frac{v}{c}A = \eta\frac{r\omega}{c}2\pi rL = \frac{2\pi\eta r^2\omega L}{c} \qquad [12]$$

The torque T is thus:

$$T = Fr = \frac{2\pi\eta r^3 \omega L}{c} \qquad\qquad [13]$$

Hence the power needed to overcome the viscous resistance is:

$$\text{power} = T\omega = \frac{2\pi\eta r^3 \omega^2 L}{c} \qquad\qquad [14]$$

Example

A journal bearing of 50 mm diameter and length 90 mm runs at 15 rev/s. Calculate the power needed to overcome viscous resistance if it is lubricated with oil of dynamic viscosity 0.05 N s/m² and there is a radial clearance of 0.02 mm.

Using equation [14]:

$$\begin{aligned}
\text{power} &= \frac{2\pi\eta r^3 \omega^2 L}{c} \\[2mm]
&= \frac{2\pi \times 0.05 \times 0.025^3 \times (2\pi \times 15)^2 \times 0.09}{0.00002} \\[2mm]
&= 196 \text{ W}
\end{aligned}$$

Revision

5   A journal bearing of 50 mm diameter and length 60 mm runs at 20 rev/s. Calculate the power needed to overcome viscous resistance if it is lubricated with oil of dynamic viscosity 0.1 N s/m² and there is a radial clearance of 0.2 mm.

7.4 Laminar and turbulent flow

When a fluid flows over a solid surface, the fluid immediately in contact with the surface, i.e. the boundary of the flow, is held at rest as a consequence of intermolecular forces. Viscous forces then result in a velocity gradient. Therefore close to the boundary there is a difference in velocity between the fluid and fluid further away (Figure 7.14). The region in which this velocity gradient occurs is called the *boundary layer*. Outside the boundary layer the velocity is unaffected by the presence of the boundary.

The flow of a fluid over a solid surface may be either laminar or turbulent. With *laminar flow*, layers of fluid can be considered as sliding smoothly over each other. The result is a boundary layer with a velocity gradient. With *turbulent flow*, the layers of fluid no longer flow smoothly over each other but mix in a complicated manner due to the presence of small vortices and eddies. As with the laminar flow, when a fluid flows over a solid surface the layer of fluid immediately adjacent to the surface is held at rest. We thus start off with a velocity gradient for the fluid very close to the boundary. However, the turbulent nature of the flow results in molecules becoming much more mixed than with laminar flow and so

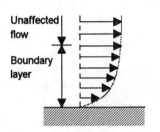

Figure 7.14   *Flow in a boundary layer*

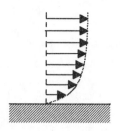

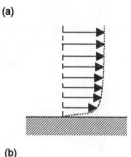

(a)

(b)

Figure 7.15   *Flow: (a)
laminar, (b) turbulent*

the fluid more rapidly comes to its common rate of flow. Figure 7.15 shows how typically the fluid velocity varies with distance from a boundary for (a) laminar flow, (b) turbulent flow. The turbulent nature of the flow results in continuously fluctuating velocities in the direction of the flow; thus the arrows shown for the velocities with the turbulent flow represent the average values over a period of time.

7.4.1 Reynolds number

For flow through a pipe, Reynolds defined an index number which if exceeded will generally result in turbulent flow but if not exceeded will give laminar flow. This index is known as the *Reynolds number*, being denoted in equations by the symbol Re. It is defined as:

$$\text{Reynolds number } Re = \frac{\rho v d}{\eta} \qquad [15]$$

where ρ is the fluid density, v the fluid velocity, d the diameter of the tube and η the fluid dynamic viscosity. When the value of the Reynolds number is greater than about 2000 then turbulent flow is likely, when less than 2000 laminar flow.

Example

Will the flow of oil through a pipe of diameter 100 mm be likely to be laminar or turbulent if it has a velocity of 0.06 m/s? The oil has a dynamic viscosity of 0.0015 N s/m^2 and a density of 750 kg/m^3.

Using equation [15]:

$$\text{Reynolds number} = \frac{\rho v d}{\eta} = \frac{750 \times 0.06 \times 0.100}{0.0015} = 3000$$

Because this is greater than 2000 the flow is likely to be turbulent.

Revision

6   What will be the critical velocity at which the transition from laminar to turbulent flow is likely for water flowing through a pipe of diameter 100 mm? Water has a density of 1000 kg/m^3 and a dynamic viscosity of 0.001 N s/m^2.

7.5 Energy loss with fluids flowing through pipes

When a fluid flows through a pipe, energy losses occur. These can arise from sudden changes in the pipe diameter, bends, elbows, etc. and from the viscous drag between the fluid and the pipe walls. For constant diameter long straight lengths of pipe, we need only consider the energy losses from the viscous drag forces with the pipe walls.

Consider the flow of a fluid through a horizontal pipe of diameter d and length L (Figure 7.16) when the input pressure is p_1 and the output

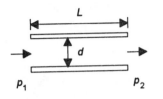

Figure 7.16   *Flow through a pipe*

pressure p_2. The force acting on the fluid in the pipe due to the pressure difference between the ends is:

$$\text{force from pressure} = p_1 \times \tfrac{1}{4}\pi d^2 - p_2 \times \tfrac{1}{4}\pi d^2 \qquad [16]$$

If the fluid is flowing through the pipe with a constant velocity, i.e. no acceleration, there can be no net force acting on the fluid. Thus the force arising from the pressure drop must be balanced by the viscous drag forces. For flow through pipes, Reynolds found the viscous drag force to be proportional to v^n, where v is the fluid velocity and n is a factor which has a value of about 1 for laminar flow and 2 for turbulent flow. Note that the fluid velocity referred to here is the mean velocity of the flow through the pipe and is the flow rate Q divided by the cross-sectional area A, i.e. $v = Q/A$. The viscous drag force also depends on the tube area A concerned. Thus:

$$\text{viscous drag force} \propto A v^n$$

This is written as:

$$\text{viscous drag force} = \tfrac{1}{2}\rho f A v^n \qquad [17]$$

where ρ is the fluid density and f is the *friction coefficient*. For most purposes n is assumed to be 2 (see Section 7.5.1) and thus:

$$\text{viscous drag force} = \tfrac{1}{2}\rho f A v^2 \qquad [18]$$

Thus, using equation [18] with $A = \pi d L$, then:

$$\text{viscous drag force} = \tfrac{1}{2}\rho f \pi d L v^2 \qquad [19]$$

Equating the viscous drag force with the force arising from the pressure drop (equation [16]) gives:

$$p_1 \times \tfrac{1}{4}\pi d^2 - p_2 \times \tfrac{1}{4}\pi d^2 = \tfrac{1}{2}\rho f \pi d L v^2$$

and thus the pressure drop is:

$$\text{pressure drop} = p_1 - p_2 = \frac{2\rho f L v^2}{d} \qquad [20]$$

This equation is often written in terms of the pressure head. Thus since $p_1 = h_1 \rho g$ and $p_2 = h_2 \rho g$, where ρ is the fluid density, h_1 the height of fluid which the pressure would support at the inlet to the tube and h_2 the height at the outlet:

$$\text{loss of head} = h_1 - h_2 = \frac{2 f L v^2}{g d} \qquad [21]$$

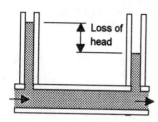

Figure 7.17   *Loss of head*

This is known as *Darcy's formula*. Note that it is often written as:

$$\text{loss of head} = \frac{4fL}{d}\frac{v^2}{2g}$$ [22]

since $v^2/2g$ is the kinetic energy of the fluid per unit mass. For the flow of liquid through a tube we can have the arrangement shown in Figure 7.17 and the loss of head is the difference in the heights of the liquid in the vertical tubes at the inlet and outlet.

Example

Water is supplied through a pipe of diameter 200 mm and length 100 m at 0.1 m³/s. Determine the head loss over the pipe length if the friction coefficient is 0.005.

The mean flow velocity v is Q/A and is thus:

$$v = \frac{0.1}{\pi \times 0.100^2} = 3.18 \text{ m/s}$$

Using Darcy's formula (equation [21]):

$$\text{loss of head} = \frac{2fLv^2}{gd} = \frac{2 \times 0.005 \times 100 \times 3.18^2}{9.81 \times 0.200} = 5.15$$

Revision

7   Determine the head loss when oil is pumped through a pipeline of diameter 100 mm and length 1.6 km at the rate of 50 m³/h. The frictional coefficient is 0.15.

7.5.1 The friction coefficient

The experiments of Reynolds gave the viscous drag for flow through a pipe as proportional to v^n with n having the value 1 for laminar flow and 2 for turbulent flow. This is expressed in terms of a frictional coefficient f by equation [17] as:

$$\text{viscous drag force} = \tfrac{1}{2}\rho fAv^n$$ [23]

where ρ is the fluid density. For the derivation of Darcy's equation n was taken to be 2. This would appear to suggest that Darcy's equation is only valid for turbulent flow. However, we can take account of this by taking the frictional coefficient for laminar flow to be proportional to $1/v$ and so Darcy's formula can also be used for laminar flow. For laminar flow it can be shown that:

$$f = \frac{16\eta}{\rho v d}$$ [24]

and since Reynolds number = $\rho v d/\eta$ (equation [15]), then:

$$f = \frac{16}{\text{Reynold's number}} \qquad [25]$$

For turbulent flow through smooth pipes it is found that:

$$f = \frac{0.079}{(\text{Reynold's number})^{1/4}} \qquad [26]$$

For rough pipes the surface texture affects the value of the friction coefficient.

Example

Oil is pumped at the rate of 50×10^3 kg/h along a pipeline of diameter 100 mm. Determine the frictional coefficient if the oil has a density of 915 kg/m^3 and a dynamic viscosity of 1.7 N s/m^2.

The mean flow velocity v is $Q/A = (50 \times 10^3/915 \times 3600)/\pi \times 0.050^2$ = 1.93 m/s. The Reynolds number for the flow is given by equation [15] as:

$$\text{Reynolds number} = \frac{\rho v d}{\eta} = \frac{915 \times 1.93 \times 0.100}{1.7} = 103.$$

This is below 2000 and so the flow is laminar. Thus, using equation [25]:

$$f = \frac{16}{103.9} = 0.15$$

Revision

8 Water flows along a pipe of diameter 50 mm and length 6 m at the rate of 0.3 m^3/s. Determine whether the flow is laminar or turbulent and the pressure drop over that length if water has a density of 1000 kg/m^3 and a dynamic viscosity of 0.0012 N s/m^2.

7.5.2 Power transmission through pipelines

Consider a horizontal pipe. The potential energy of the fluid, with reference to the level of the pipe, at the inlet to a horizontal pipe is mgh_1, where h_1 is the height of the column of fluid supported by the fluid pressure at that point. If the volume of fluid entering the pipe per second is Q, then the mass m of fluid entering the pipe per second is $Q\rho$, where ρ is the fluid density. Thus the power supplied at the pipe inlet is $Q\rho g h_1$. Likewise, the power available at the pipe outlet is $Q\rho g h_2$, where h_2 is the height of the column of fluid supported by the fluid pressure at that point. If h_f is the loss of head for flow through the pipe, then $h_2 = h_1 - h_f$. Thus the power transmitted is:

$$\text{power transmitted } P = Q\rho g(h_1 - h_f) \qquad [27]$$

Since $Q = Av$, where v is the mean velocity and A the cross-sectional area of the tube, and using Darcy's formula (equation [21]), then equation [27] can be written as:

$$\text{power transmitted } P = \rho g A v \left(h_1 - \frac{2fLv^2}{gd} \right) \qquad [28]$$

The power transmitted will be a maximum when $dP/dv = 0$, thus writing equation [28] as:

$$P = \rho g A \left(h_1 v - \frac{2fLv^3}{gd} \right)$$

then:

$$\frac{dP}{dv} = \rho g A \left(h_1 - \frac{3 \times 2fLv^2}{gd} \right)$$

Hence the maximum power occurs when:

$$h_1 = \frac{3 \times 2fLv^2}{gd} = 3h_f \qquad [29]$$

The power transmitted is a maximum when the flow is such that one third of the total head is lost in overcoming friction.

Example

A pump supplies water at a pressure of 5 MPa to one end of a pipeline. If the pipeline has a length of 1 km and a diameter of 200 mm, what is the maximum available power that can be delivered at the other end of the pipe. Take the frictional coefficient to be 0.005 and the density of water as 1000 kg/m³.

The input pressure head is $h_1 = p/\rho g = 5 \times 10^6/1000 \times 9.81 = 510$ m. For maximum power transmission $h_f = h_1/3$ (equation [29]) and so the head lost $h_f = 510/3 = 170$ m. Hence, using Darcy's formula (equation [21]):

$$h_f = \frac{2fLv^2}{gd}$$

then:

$$v^2 = \frac{h_f g d}{2fL} = \frac{170 \times 9.81 \times 0.200}{2 \times 0.005 \times 1000}$$

and $v = 5.78$ m/s. The power available is given by equation [27] as:

$$\text{power available} = Q\rho g(h_1 - h_f) = Q\rho g(h_1 - h_1/3) = 2Q\rho gh_1/3$$

Since $Q = Av = \pi r^2 v$, then:

$$\text{power available} = 2\pi r^2 v\rho gh_1/3 = 2\pi r^2 vp/3$$

$$= 2\pi \times 0.100^2 \times 5.78 \times 5 \times 10^6/3 = 605 \text{ kW}$$

Example

Oil is pumped though a horizontal pipeline of 150 mm diameter and length 1 km at the rate of 100 kg/s. If the oil has a density of 850 kg/m^3 and a dynamic viscosity of 0.12 N s/m^2, determine the power required.

The volume rate of flow Q is $100/850 = 0.118$ m^3/s and so the mean velocity of the flow is $Q/A = 0.118/\pi \times 0.075^2 = 6.68$ m/s. Reynolds number for the flow is given by equation [15] as:

$$\text{Reynolds number} = \frac{\rho vd}{\eta} = \frac{850 \times 6.68 \times 0.150}{0.12} = 7097.$$

The flow is thus turbulent and so the friction coefficient is given by equation [26] as:

$$f = \frac{0.079}{(\text{Reynold's number})^{1/4}} = \frac{0.079}{7097.5^{1/4}} = 0.0086$$

Using Darcy's formula (equation [21]), the head loss h_f is:

$$h_f = \frac{2fLv^2}{gd} = \frac{2 \times 0.0086 \times 1000 \times 6.68^2}{9.81 \times 0.150} = 521.6$$

The power P required to overcome this head loss is mgh_f, where m is the rate of mass flow through the pipe. thus:

$$P = 100 \times 9.81 \times 521.6 = 511.7 \text{ kW}$$

Revision

9   Determine the head necessary to pump water through a pipe of diameter 300 mm and length 1 km if the mean velocity required is 2.5 m/s and the transmitted power 200 kW. Take the frictional coefficient to be 0.005 and the density of water 1000 kg/m^3.

10   Determine the power required to pump oil through a pipe of diameter 100 mm and length 1.0 km at 50 000 kg/hour if the oil has a density of 915 kg/m^3 and a dynamic viscosity of 1.7 N s/m^2.

Problems

1   The dynamic viscosity of water at 20°C is 1.00×10^{-3} N s/m^2 and its density is 1000 kg/m^3. What is its kinematic viscosity at this temperature?

2   A solid circular cylinder of diameter 100 mm, length 200 mm and weight 20 N, slides inside a vertical concentric pipe of internal diameter 100.5 mm, the space between the two being an oil film. If the oil has a dynamic viscosity of 0.3 N s/m^2, what will be the descent velocity of the cylinder?

3   Two plates are separated by a 0.2 mm layer of lubricant. The lower plate is stationary and the upper plate moves with a constant velocity of 5 m/s. Determine the shear stress in the oil if it has a dynamic viscosity of 0.2 N s/m^2.

4   A shaft of diameter 80.0 mm is being pushed with a constant velocity of 0.4 m/s through a concentric bearing sleeve of diameter 80.2 mm and 200 mm long. The space between the shaft and the cylinder is filled with a lubricant of dynamic viscosity 0.15 N s/m^2. Determine the resistant force and the power required.

5   A collar thrust bearing is used with a shaft rotating at 1200 rev/min and with the load transmitted to an annular surface of internal diameter 60 mm and external diameter 120 mm via a lubricant layer of thickness 0.2 mm and dynamic viscosity 0.1 N s/m^2. Determine the power required to overcome the viscous torque exerted by the lubricant.

6   A vertical shaft of radius 40 mm rotates at 15 rev/s in a footstep bearing, the thrust being transmitted through a film of lubricant of thickness 0.2 mm and dynamic viscosity 0.1 N s/m^2. Determine the power required to overcome the viscous resistance.

7   A journal bearing of 70 mm diameter and length 150 mm runs at 1000 rev/min. Calculate the power needed to overcome viscous resistance if it is lubricated with oil of dynamic viscosity 0.2 N s/m^2 and there is a radial clearance of 0.25 mm.

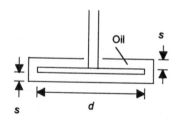

Figure 7.18   *Problem 8*

8   Instruments having pointers which have an angular motion are sometimes damped by means of a disc connected to their shaft, the disc being immersed in a container of oil as shown in Figure 7.18. Derive an equation for the damping torque as a function of the disc diameter d, the spacing s of the disc from the container surfaces, the rate of angular rotation ω and the dynamic viscosity η of the oil.

9   A journal bearing of 50 mm diameter and length 75 mm runs at 2000 rev/min. Calculate the power needed to overcome viscous resistance if it is lubricated with oil of dynamic viscosity 0.2 N s/m^2 and there is a radial clearance of 0.15 mm.

10 What will be the critical velocity at which the transition from laminar to turbulent flow is likely for SAE 30 oil flowing through a pipe of diameter 50 mm? The oil has a density of 880 kg/m^3 and a dynamic viscosity of 0.3 N s/m^2.

11 Determine the head loss when oil is pumped through a pipeline of diameter 150 mm and length 0.5 km at the rate of 25 kg/s. The frictional coefficient is 0.01 and the oil has a density of 850 kg/m^3.

12 Determine whether the flow is laminar or turbulent and the head loss when oil is pumped through a pipeline of diameter 150 mm and length 0.5 km at the rate of 25 kg/s. The oil has a density of 850 kg/m^3 and a dynamic viscosity of 0.12 N s/m^2.

13 Determine whether the flow is laminar or turbulent and the head loss when oil is pumped through a pipeline of diameter 25 mm and length 3 m at a mean velocity of 0.32 m/s. The oil has a density of 900 kg/m^3 and a dynamic viscosity of 1.44 N s/m^2.

14 Oil is pumped though a horizontal pipeline of 75 mm diameter and length 0.9 km at the rate of 40 000 kg/h. If the oil has a density of 850 kg/m^3 and a dynamic viscosity of 0.28 N s/m^2, determine the power required.

15 Water is pumped though a horizontal pipeline of 6 mm diameter and length 100 m at a mean velocity of 0.2 m/s. If the water has a density of 1000 kg/m^3 and a dynamic viscosity of 0.0013 N s/m^2, determine the power required.

16 Oil is pumped though a horizontal pipeline of 75 mm diameter and length 1 km at the rate of 2.75 kg/s. If the oil has a density of 900 kg/m^3 and a dynamic viscosity of 0.17 N s/m^2, determine the power required.

8 Single phase a.c. theory

8.1 Introduction

This chapter reviews the basic principles of a.c. theory and uses them to solve problems involving circuits containing resistance, inductance and capacitance in series and in parallel. The approach is in a general manner involving the consideration of phasors and their addition by means of 'phasor diagrams'; Chapter 9 then uses complex numbers to solve such problems.

8.1.1 Sinusoidal quantities and phasors

We can represent a voltage v which varies sinusoidally with time t by the equation $v = V \sin \omega t$, where V is the maximum value of the voltage and ω the angular frequency and equal to $2\pi f$. We can imagine such a signal being produced by the vertical projection of a radial line of length V rotating with a constant angular velocity ω from some initial start position (Figure 8.1). Thus instead of specifying the variation of the voltage with time by the above equation, we can specify it by the length of the line V and whether it starts at $t = 0$ at some angle, termed the *phase angle* ϕ, to the reference axis which is usually taken as the horizontal axis. This specification can be written as $V \angle \phi$. Such lines are termed *phasors*.

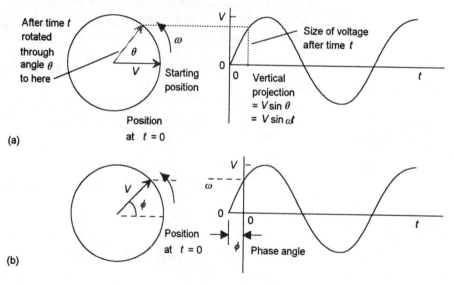

Figure 8.1 *(a)* $v = V \sin \omega t$, *(b)* $v = V \sin(\omega t + \phi)$

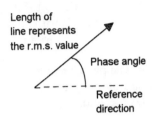

Figure 8.2   *A phasor*

A *phasor* can be described by drawing an arrow-headed line, the length of the line representing the amplitude and its direction, relative to a reference direction, as the phase angle (Figure 8.2). Because with alternating currents and voltages we are normally concerned with root-mean-square (r.m.s.) values rather than maximum value, for sinusoidal waves the maximum value is just the r.m.s. value divided by $\sqrt{2}$, generally when the term phasor is used for an arrow-headed line describing alternating currents and voltages the length of the line represents the r.m.s. value.

The phase angle is the angle between a phasor and some reference direction. In the case of a series circuit it is customary to use the direction of the current phasor for the circuit as the reference direction, the current being the same for all the series components. For a parallel circuit it is customary to use the direction of the voltage phasor for the parallel circuit as the reference phasor, the voltage being the same for all parallel components.

In textbooks the common practice to indicate that a symbol represents a phasor is to use bold print, e.g. **V** represents a voltage phasor. The voltage representing the length of the phasor would be given by the italic, non-bold, symbol *V*. The instantaneous value of the voltage is represented by a lower case *v*.

8.1.2 Addition and subtraction of phasors

Phasors can be added, or subtracted, by the methods used to add, or subtract, vector quantities. One method to add two phasors is the *parallelogram law* (the same parallelogram law as used for the addition of vector quantities). If the phasors are drawn to scale as arrow-headed lines then we draw one phasor, then draw the next phasor so that its tail is attached to the tail of the preceding one, and then complete a parallelogram with lines parallel to these two phasors; the line from junctions of the tails of the two phasors to give the diagonal of the parallelogram represents the result of the addition (Figure 8.3). Since a negative phasor –**V** is just **V** with the direction reversed, subtraction is accomplished by the addition of the negative of a phasor (Figure 8.4).

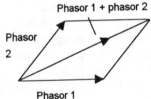

Figure 8.3   *Addition of phasors by the parallelogram law*

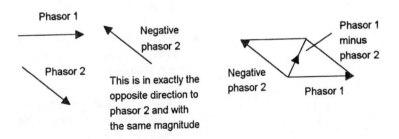

Figure 8.4 *Subtraction of one phasor from another*

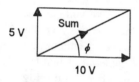

Figure 8.5　*Example*

Example

Determine the phasor to represent the voltage across a circuit if it is the sum of the phasors 10 V with phase angle 0° and 5 V with phase angle 90°, i.e. add $10\angle 0°$ V and $5\angle 90°$ V.

Figure 8.5 shows the phasor parallelogram. The phasor representing the sum is the diagonal of the parallelogram from the junction of the two tails of the phasor. Using the Pythagoras theorem,

$$(\text{diagonal})^2 = 5^2 + 10^2$$

Hence the sum phasor has a magnitude of 11.2 V. This is at a phase angle ϕ where $\tan \phi = 5/10$ and so ϕ is 26.6°. The voltage leads the 10 V voltage by 26.6°. The term 'lead' is used when the angle is an anticlockwise rotation from the reference phasor, in this case the 10 V phasor. The phasor is thus $11.2\angle 26.6°$ V.

Example

The phasor representing the current entering an arrangement of two parallel components is 4 A at zero phase angle. If the phasor representing the current through one component is 2 A at a phase angle of 90°, determine the phasor to represent the current through the other component .

We need to subtract the phasor for the current through one component from the phasor for the total current. We can do this by adding to the phasor for the total current the negative phasor for the current through the component. Figure 8.6 shows the phasors and the 'sum' phasor. The current through the second component is thus $\text{Å}(4^2 + 2^2) = 4.5$ A at a phase angle given by $\tan \phi = -2/4$ and thus $\phi = -25.6°$. The current lags the overall current by 25.6°. the term 'lag' is used when the current phasor is at a phase which involves a clockwise rotation from the reference phasor.

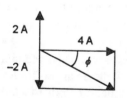

Figure 8.6　*Example*

Revision

1　Determine the phasor to represent the voltage across a circuit if it is the sum of the phasors 5 V with phase angle 0° and 2 V with phase angle 90°.

2　Determine the phasor to represent the current leaving a parallel circuit if it is the sum of the phasors 2 A with phase angle 0° and 3 A with phase angle 90°.

3　If the voltage across two series components is represented by a 10 V phasor with zero phase angle and the voltage phasor across one of them is 4 V at 90° phase angle, determine the voltage phasor for the other component.

8.2 Reactance and susceptance

The *reactance X* of a component is the maximum voltage V_m across the component divided by the maximum current I_m through it and has the unit of ohms. Note that these maximum values may not occur at the same time.

$$X = \frac{V_m}{I_m} \qquad [1]$$

Since, with sinusoidal signals, the root-mean-square values $V_{r.m.s} = V_m/\sqrt{2}$ and $I_{r.m.s} = I_m/\sqrt{2}$, we can also write:

$$X = \frac{V_{r.m.s}}{I_{r.m.s}} \qquad [2]$$

With d.c. we have the term conductance for the reciprocal of resistance; with a.c. the term *susceptance B* is used for the reciprocal of reactance X and has the unit of $/\Omega$ which is given the name siemen (S).

$$B = \frac{1}{X} \qquad [3]$$

Example

What are the reactance and susceptance of a component if the alternating voltage drop across it has a root-mean-square value of 20 V when the root-mean- square current through it is 50 mA.

The reactance X is $V_{r.m.s}/I_{r.m.s.} = 20/0.050 = 400 \ \Omega$. The susceptance is the reciprocal of reactance and so is $1/400 = 0.0025$ S.

Revision

4   What are the reactance and susceptance of a component if the alternating voltage drop across it has a root-mean-square value of 5 V when the root-mean-square current through it is 2 mA?

8.3 Phasor relationships for pure components

The term 'pure' is used since it is assumed that the components only have the single property concerned. Thus a pure inductor is assumed to have only inductance and no resistance or capacitance.

1   *Pure resistor*
    For a pure resistor R the voltage drop V across it is $v = Ri$ and thus for $i = I \sin \omega t$ we have $v = Ri \sin \omega t$ and so the voltage is in phase with the current (Figure 8.7(a)). The maximum value of the current occurs at the same time as the maximum value of the voltage and the ratio of the maximum voltage to the maximum current or the r.m.s. voltage to the r.m.s. current is the resistance.

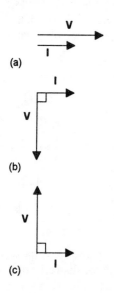

(a)

(b)

(c)

Figure 8.7   *Phasors for:*

(a) resistors, (b) capacitors,

(c) inductors

2   *Pure capacitor*

For a pure capacitor of capacitance C the charge $q = Cv$ and thus, for $v = V \sin \omega t$, we have $q = CV \sin \omega t$ and so the current $i = dq/dt = \omega CV \cos \omega t = \omega CV \sin(\omega t + 90°)$. The current leads the voltage across the capacitor by 90° (Figure 8.7(b)). Alternatively we can say that the voltage lags the current by 90°. Thus the maximum voltage and maximum current do not occur at the same time. The ratio of the maximum voltage to maximum current or the r.m.s. voltage to r.m.s. current, i.e. the capacitive reactance X_C, is:

$$X_C = \frac{1}{2\pi f C} = \frac{1}{\omega C} \qquad\qquad [4]$$

3   *Pure inductor*

For a pure inductor of inductance L the voltage drop across it $v = L \, di/dt$ and thus for a current of $i = I \sin \omega t$ we have $v = \omega LI \cos \omega t = \omega LI \sin (\omega t + 90°)$. The voltage leads the current by 90° or alternatively we can say the current lags the voltage by 90° (Figure 8.7(c)). Thus the maximum voltage and the maximum current do not occur at the same time. The ratio of the maximum voltage to maximum current or the r.m.s. voltage to r.m.s. current, i.e. the inductive reactance X_L, is:

$$X_L = 2\pi f L = \omega L \qquad\qquad [5]$$

A useful way of remembering the phase relationships with capacitors and inductors is the mnemonic:

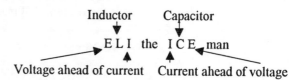

Inductor   Capacitor

E L I   the   I C E   man

Voltage ahead of current   Current ahead of voltage

Another mnemonic that can be used is:

In C, current leads voltage

C I V I L

Voltage leads current in L

Example

Determine the reactance of a pure inductor if it has an inductance of 200 mH and is used in a circuit where the frequency of the alternating current is 50 Hz. What will be the r.m.s. current through the inductor when the r.m.s. voltage drop across it is 20 V?

The reactance is given by equation [5] as:

$$X_L = 2\pi fL = 2\pi \times 50 \times 0.200 = 62.8 \ \Omega$$

The r.m.s. current $I_{r.m.s}$ is given by:

$$I_{r.m.s} = \frac{V_{r.m.s}}{X_L} = \frac{20}{62.8} = 0.32 \text{ A}$$

Example

If the alternating current through a 20 μF capacitor is 3 sin 800*t* A, what is the voltage across it?

With ω = 800 rad/s, the reactance of the capacitor is given by equation [4] as:

$$X_C = \frac{1}{\omega C} = \frac{1}{800 \times 20 \times 10^{-6}} = 62.5 \ \Omega$$

The maximum value of the current is 3 A. Hence the maximum value of the voltage is $V = IX_C = 3 \times 62.5 = 187.5$ V. The voltage is thus 187.5 sin 800*t* V.

Revision

5   Determine the reactance of a pure capacitor if it has a capacitance of 8 μF and is used in a circuit where the frequency of the alternating current is 1 kHz. What will be the r.m.s. current through the capacitor when the r.m.s. voltage drop across it is 10 V?

6   Determine the capacitive reactance of a 0.5 μF capacitor, and the current through it, when the voltage across it is 16 sin 2000*t* V.

7   Determine the inductive reactance of a pure inductor of inductance 80 mH when the current through it is 100 sin 400*t* mA and the voltage drop across it.

8.4 Impedance and admittance

In discussing series and parallel a.c. circuits involving resistors, capacitors and inductors, the term impedance will be used.

> The *impedance* Z of a circuit is the phasor value of the voltage across the circuit divided by the phasor value of the current through it.

Impedance is not a sinusoidally varying quantity and thus is not a phasor, so bold print is not used for it in this book (though some textbooks use bold print because it is a complex quantity). Impedance has the unit of ohms.

$$Z = \frac{\mathbf{V}}{\mathbf{I}} \qquad\qquad [6]$$

Admittance Y is the reciprocal of impedance Z. Impedance is a measure of how well a component impedes the current, admittance is a measure of how it admits, i.e. allows, the current.

$$Y = \frac{1}{Z} \qquad\qquad [7]$$

The unit of admittance is $/\Omega$ or siemen (S). Thus an impedance of 10 Ω is an admittance of $1/10 = 0.1$ S.

8.5 Series a.c. circuits

For a series d.c. circuit there is the same current through each component and the voltage drop across all the components is the sum of the voltage drops across each. For a series a.c. circuit there is the same current phasor for the current through each component and the phasor for the voltage drop across all the components is the sum of the phasors for the voltage drops across each component.

1   *Series circuit containing resistance and inductance*
For such a circuit (Figure 8.8(a)), the voltage for the resistance is in phase with the current and the voltage for the inductor leads the current by 90° (Figure 8.8(b)). Thus the phasor for the sum of the voltage drops across the two series components is given by Figure 18.7(c) as a voltage phasor **V** with a phase angle ϕ. We can use the Pythagoras theorem to give the magnitude V of the voltage:

$$V^2 = V_R^2 + V_L^2 \qquad\qquad [8]$$

and trigonometry to give the phase angle ϕ, i.e. the angle by which the voltage leads the current, which is in the direction of $\mathbf{V_R}$:

$$\tan\phi = \frac{V_L}{V_R} \qquad\qquad [9]$$

or

$$\cos\phi = \frac{V_R}{V} \qquad\qquad [10]$$

Since $V_R = IR$ and $V_L = IX_L$ and the magnitude V of **V** is given by $V = IZ$, then by substitution in equations [8], [9] and [10]:

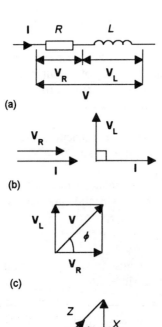

(a)

(b)

(c)

(d)

Figure 8.8   *RL series circuit*

$$Z^2 = R^2 + X_L^2 \qquad [11]$$

$$\tan\phi = \frac{X_L}{R} \qquad [12]$$

$$\cos\phi = \frac{R}{Z} \qquad [13]$$

Alternatively, we can arrive at the above equations taking the voltage triangle formed by half the parallelogram in Figure 8.8(c) and, since the voltages are IR, IX_L and IZ, if we divide each side by I we end up with an impedance triangle shown in Figure 8.8(d) The values of the impedance and the phase angle can then be determined from the impedance triangle by the use of the Pythagoras theorem and trigonometry:

Example

In a series RL circuit, the resistance is 10 Ω and the inductance 50 mH. Determine the value of the current and its phase angle with respect to the voltage if a 10 V r.m.s., 50 Hz supply is connected to the circuit.

The inductive reactance is $X_L = 2\pi f L = 2\pi \times 50 \times 0.05 = 15.7\ \Omega$. Thus equations [11] and [12] give:

$$Z = \sqrt{R^2 + X_L^2} = \sqrt{10^2 + 15.7^2} = 18.6$$

$$\phi = \tan^{-1}\frac{X_L}{R} = \tan^{-1}\frac{15.7}{10} = 57.5°$$

Hence the magnitude of the current I is given by:

$$I = \frac{V}{Z} = \frac{10}{18.6} = 0.54\ \text{A}$$

and, since the current is in the direction of the phasor $\mathbf{V_R}$, its phase angle is 57.5° lagging behind the applied voltage $\mathbf{V}$.

2 *Series circuit containing resistance and capacitance*
For such a circuit (Figure 8.9(a)), the voltage for the resistance is in phase with the current and the voltage for the capacitor lags the current by 90° (Figure 8.9(b)). Thus the phasor for the sum of the voltage drops across the two series components is given by Figure 8.9(c) as a voltage phasor with a phase angle ϕ. We can use the Pythagoras theorem to give the magnitude V of the voltage:

$$V^2 = V_R^2 + V_C^2 \qquad [14]$$

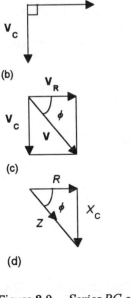

(a)

(b)

(c)

(d)

Figure 8.9 *Series RC circuit*

and trigonometry to give the phase angle ϕ, i.e. the angle by which the current leads the voltage:

$$\tan \phi = \frac{V_C}{V_R} \qquad [15]$$

$$\cos \phi = \frac{V_R}{V} \qquad [16]$$

Since $V_R = IR$ and $V_C = IX_C$ and the magnitude of $\mathbf{V}$ is I multiplied by the magnitude of the impedance, i.e. $V = IZ$, by substitution in equations [14], [15] and [16] we obtain;

$$Z^2 = R^2 + X_C^2 \qquad [17]$$

$$\tan \phi = \frac{X_C}{R} \qquad [18]$$

$$\cos \phi = \frac{R}{Z} \qquad [19]$$

Alternatively as the voltage parallelogram in Figure 8.9(c) has sides of lengths IR, IX_C and IZ, if we divide each side by I we end up with an impedance triangle shown in Figure 8.9(d) The above equations for the impedance and the phase angle then can be determined by the use of the Pythagoras theorem and trigonometry:

Example

In a series RC circuit the resistance is 20 Ω and the capacitance 50 µF. Determine the value of the current and its phase angle with respect to the voltage if a 240 V r.m.s., 50 Hz supply is connected to the circuit.

The capacitive reactance is $1/2\pi fC = 1/(2\pi \times 50 \times 50 \times 10^{-6}) = 63.7\ \Omega$. Thus equations [17] and [18] give:

$$Z = \sqrt{R^2 + X_C^2} = \sqrt{20^2 + 63.7^2} = 66.8\ \Omega$$

$$\phi = \tan^{-1}\frac{X_C}{R} = \tan^{-1}\frac{63.7}{20} = 72.6°$$

Hence the magnitude of the current I is given by:

$$I = \frac{V}{Z} = \frac{240}{66.8} = 3.59\ \text{A}$$

and, since the current is in the direction of the phasor $\mathbf{V_R}$, its phase angle is 72.6° leading the applied voltage $\mathbf{V}$.

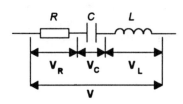

Figure 8.10   *Series RCL circuit*

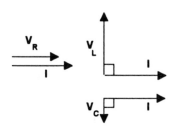

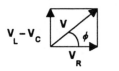

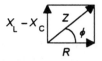

Figure 8.11 $V_L > V_C$

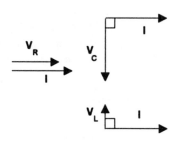

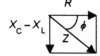

Figure 8.12 $V_L < V_C$

3   *Series circuit containing resistance, capacitance and inductance*

For such a circuit (Figure 8.10), the voltage across the resistance is in phase with the current, the voltage across the capacitor lags the current by 90° and the voltage across the inductor leads the current by 90°.

For $V_L > V_C$, i.e. $X_L < X_C$ (Figure 8.11): because the voltage phasors for the inductor and capacitor are in opposite directions we can subtract them to give a phasor for the voltage drop across the inductor and capacitor of $V_L - V_C$. Then, for the voltage triangle:

$$V^2 = V_R^2 + (V_L - V_C)^2 \qquad [20]$$

$$\tan\phi = \frac{V_L - V_C}{V_R} \qquad [21]$$

$$\cos\phi = \frac{V_R}{V} \qquad [22]$$

As a result the circuit behaves as though it was inductance in series with resistance, the voltage across the series arrangement leading the current by ϕ. Since $V_R = IR$, $V_L = IX_L$ and $V_C = IX_C$ and the magnitude of **V** is I multiplied by the magnitude of the impedance, i.e. $V = IZ$, by substitution for the voltages in equations [14], [15] and [16]; or alternatively by constructing the impedance triangle:

$$Z^2 = R^2 + (X_L - X_C)^2 \qquad [23]$$

$$\tan\phi = \frac{X_L - X_C}{R} \qquad [24]$$

$$\cos\phi = \frac{R}{Z} \qquad [25]$$

For $V_L < V_C$, i.e. $X_L < X_C$ (Figure 8.12): because the voltage phasors for the inductor and capacitor are in opposite directions we can subtract them to give a phasor for the voltage drop across the inductor and capacitor of $V_C - V_L$. For the voltage triangle:

$$V^2 = V_R^2 + (V_C - V_L)^2 \qquad [26]$$

$$\tan\phi = \frac{V_C - V_L}{V_R} \qquad [27]$$

$$\cos\phi = \frac{V_R}{V} \qquad [28]$$

As a result the circuit behaves as though it was capacitance in series with resistance, the voltage across the series arrangement lagging behind the current by ϕ. Since $V_R = IR$, $V_L = IX_L$ and $V_C = IX_C$ and the magnitude of $\mathbf{V}$ is I multiplied by the magnitude of the impedance, i.e. $V = IZ$, by substitution for the voltages in equations [26]], [27] and [28]; or alternatively by constructing the impedance triangle:

$$Z^2 = R^2 + (X_C - X_L)^2 \qquad [29]$$

$$\tan\phi = \frac{X_C - X_L}{R} \qquad [30]$$

$$\cos\phi = \frac{R}{Z} \qquad [31]$$

For $V_L = V_C$, i.e. $X_L = X_C$ (Figure 8.13): the voltage phasor for the capacitor is equal in magnitude to the voltage phasor for the inductor but in exactly the opposite direction. Thus the two voltage phasors when added cancel each other out. The total voltage is thus just $\mathbf{V_R}$. The result is that the circuit behaves as though it was just the resistance with the impedance $Z = R$ and $\phi = 0°$.

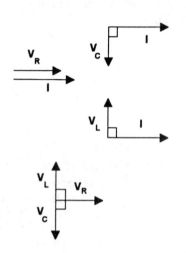

Figure 8.13 $V_L = V_C$

Example

In a series RLC circuit the resistance is 10 Ω, the inductance 60 mH and the capacitance 300 μF. Determine the value of the current and its phase angle with respect to the voltage if a 24 V r.m.s., 50 Hz supply is connected to the circuit.

The inductive reactance is $X_L = 2\pi f L = 2\pi \times 50 \times 0.060 = 18.8 \ \Omega$ and the capacitive reactance is $X_C = 1/(2\pi f C) = 1/(2\pi \times 50 \times 300 \times 10^{-6}) = 10.6 \ \Omega$. Since the inductive reactance is greater than the capacitive reactance the situation is like that shown in Figure 8.11. The circuit impedance Z is:

$$Z = \sqrt{R^2 + (X_L - X_C)^2} = \sqrt{10^2 + (18.8 - 10.6)^2} = 12.9$$

$$\phi = \tan^{-1}\frac{X_L - X_C}{R} = \tan^{-1}\frac{18.8 - 10.6}{10} = 39.4°$$

The circuit current has a magnitude I of:

$$I = \frac{V}{Z} = \frac{24}{12.9} = 1.86 \ \text{A}$$

and because the circuit behaves as an inductive load the current lags the voltage by 39.6°.

Revision

8   Determine the total impedance, and the phase angle between the overall voltage and the current, of a circuit containing a resistance of 300 Ω in series with an inductance of 200 mH when a voltage of 10 sin 2000*t* V is applied.

9   A voltage of root-mean-square value 30 V at an angular frequency of 1000 rad/s is applied to a circuit consisting of a resistance of 200 Ω in series with an inductance of 100 mH. Determine the current.

10 Determine the total impedance, and the phase angle between the overall voltage and the current, of a circuit containing a resistance of 3.3 kΩ in series with a capacitance of 2.2 μF when a voltage of 10 sin 240*t* V is applied.

11 A voltage of root-mean-square value 12 V at an angular frequency of 120 rad/s is applied to a circuit consisting of a resistance of 200 Ω in series with a capacitance of 2 μF. Determine the current.

12 A series a.c. circuit consists of a resistance of 100 Ω, a capacitance of 8 μF and an inductance of 100 mH. What is the total circuit impedance and the phase angle between the overall voltage and the current when the angular frequency is 1000 rad/s?

8.6 Parallel circuits

For a parallel d.c. circuit there is the same voltage drop across each component and the current entering the parallel arrangement is the sum of the currents through each component. For a parallel a.c. circuit there is the same voltage phasor for the voltage across each component and the phasor for the current entering the parallel arrangement is the sum of the phasors for the currents through each component.

1   *Parallel resistance and inductance*
    Figure 8.14(a) shows the parallel circuit. The voltage across each component is the same, being the supply voltage **V**. The current **I$_R$** flowing through the resistance is in phase with the supply voltage; the current **I$_L$** through the inductor lags behind the supply voltage by 90° (Figure 8.14(b)). The phasor sum of the currents **I$_R$** and **I$_L$** must be the supply current **I** entering the parallel arrangement (Figure 8.14(c)). Thus:

$$I^2 = I_R^2 + I_L^2 \qquad [32]$$

$$\tan \phi = \frac{I_L}{I_R} \qquad [33]$$

$$\cos \phi = \frac{I_R}{I} \qquad [34]$$

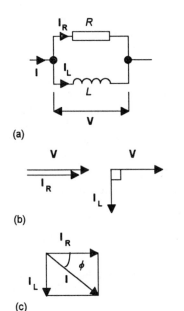

(a)

(b)

(c)

Figure 8.14   *Parallel R and L*

(a)

G

B_L

Y

(b)

Figure 8.15   *Parallel R and L*

But $V = IZ$, where Z is the impedance of the parallel arrangement and $V = I_R R$ and $V = I_L X_L$, and so by substitution in equations [32], [33] and [34]:

$$\frac{1}{Z^2} = \frac{1}{R^2} + \frac{1}{X_L^2} \qquad [35]$$

Alternatively we could redraw the current triangle in Figure 8.14(c) as Figure 8.15(a) and use the Pythagoras theorem and trigonometry to obtain the above equations.

The conductance G is the reciprocal of resistance, susceptance B the reciprocal of reactance and admittance the reciprocal of impedance. Thus, equation [35] in terms of admittance is:

$$Y^2 = G^2 + B_L^2 \qquad [36]$$

and equations [33] and [34]:

$$\tan\phi = \frac{R}{X_L} \qquad [37]$$

$$\cos\phi = \frac{Z}{R} \qquad [38]$$

Alternatively we can obtain the above equations from the impedance triangle in Figure 8.15(a); when each side length is divided by V we obtain the admittance triangle shown in Figure 8.15(b).

Example

A resistance of 10 Ω is in parallel with an inductance of 50 mH. What will be the current drawn from the supply and the impedance of the circuit when a voltage of 24 V r.m.s., 50 Hz is applied to it?

The inductive reactance is $X_L = 2\pi f L = 2\pi \times 50 \times 0.050 = 15.7\ \Omega$. Thus $I_L = V/X_L = 24/15.7 = 1.53$ A. The current through the resistor $I_R = V/R = 24/10 = 2.4$ A. The magnitude of the current drawn from the supply is thus:

$$I = \sqrt{I_R^2 + I_L^2} = \sqrt{2.4^2 + 1.53^2} = 2.85\ \text{A}$$

The phase angle ϕ by which the supply current lags the supply voltage is:

$$\phi = \tan^{-1}\frac{R}{X_L} = \tan^{-1}\frac{10}{15.7} = 32.5°$$

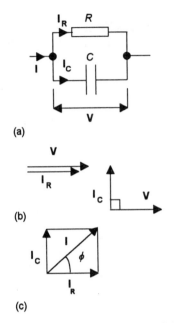

(a)

(b)

(c)

Figure 8.16   *Parallel R and C*

The circuit impedance has the value:

$$Z = \frac{V}{I} = \frac{24}{2.85} = 8.42 \ \Omega$$

2   *Parallel resistance and capacitance*

Figure 8.16(a) shows the parallel circuit. The voltage across each component is the same, being the supply voltage **V**. The current **I**R flowing through the resistance is in phase with the supply voltage; the current I_C through the inductor leads the supply voltage by 90° (Figure 8.16(b)). The phasor sum of the currents I_R and I_C is the current **I** entering the parallel arrangement (Figure 8.16(c)). Thus:

$$I^2 = I_R^2 + I_C^2 \tag{39}$$

$$\tan \phi = \frac{I_C}{I_R} \tag{40}$$

$$\cos \phi = \frac{I_R}{I} \tag{41}$$

But $V = IZ$, where Z is the impedance of the parallel arrangement and $V = I_R R$ and $V = I_C X_C$, thus by substitution in equations [39], [40] and [41]:

$$\frac{1}{Z^2} = \frac{1}{R^2} + \frac{1}{X_C^2} \tag{42}$$

Since conductance G is the reciprocal of resistance, susceptance B the reciprocal of reactance and admittance the reciprocal of impedance, equation [42] can be written as:

$$Y^2 = G^2 + B_C^2 \tag{43}$$

By substitution in equations [40] and [41]:

$$\tan \phi = \frac{R}{X_C} \tag{44}$$

$$\cos \phi = \frac{Z}{R} \tag{45}$$

Alternatively we could redraw the current triangle in Figure 8.14(c) as Figure 8.17(a). When each side length is divided by V, we obtain the admittance triangle shown in Figure 8.17(b). Hence the above equations can be obtained by the use of the Pythagoras theorem and trigonometry.

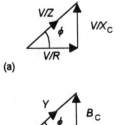

(a)

(b)

Figure 8.17   *Parallel R and C*

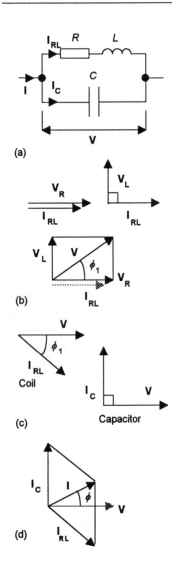

(a)

(b)

(c)

(d)

Figure 8.18   *Parallel R-L and C*

Example

A resistance of 10 Ω is in parallel with a capacitance of 10 µF. What will be the current drawn from the supply and the impedance of the circuit when a voltage of 20 V r.m.s., 1 kHz is applied to it?

The inductive capacitance is $X_C = 1/2\pi f C = 1/2\pi \times 1000 \times 10 \times 10^{-6}$ = 15.9 Ω. Thus $I_C = V/X_C = 20/15.9 = 1.26$ A. The current through the resistor $I_R = V/R = 20/10 = 2.0$ A. The magnitude of the current drawn from the supply is thus:

$$I = \sqrt{I_R^2 + I_C^2} = \sqrt{2.0^2 + 1.26^2} = 2.36 \text{ A}$$

The phase angle ϕ by which the supply current leads the voltage is:

$$\phi = \tan^{-1}\frac{R}{X_C} = \tan^{-1}\frac{10}{15.9} = 32.2°$$

The current is thus $2.36\angle 32.2°$ A. The circuit impedance has the value $Z = V/I = 20/2.36 = 8.47$ Ω.

3   *Parallel R-L and C*

A coil will have both resistance and inductance and can be regarded as resistance in series with inductance. Consider such a coil in parallel with capacitance (Figure 8.18(a)). Figure 8.18(b) shows the phasor diagram for the two series items in the coil branch. For this branch we thus have a voltage **V** which is leading $\mathbf{I_{RL}}$, the current in phase with $\mathbf{V_R}$, by a phase angle ϕ_1. We thus have the phasors for the coil branch and the capacitor branch as shown in Figure 8.18(c). The total circuit current is the phasor sum of $\mathbf{I_{RL}}$ and $\mathbf{I_C}$ (Figure 8.18(d)).

As with vector quantities, phasors can be resolved into two mutually perpendicular directions. Figure 8.18(d) shows the situation when the resolved component of $\mathbf{I_{RL}}$ in the vertical direction has a magnitude less than that of $\mathbf{I_C}$. We thus have:

$I_C > I_{RL} \sin \phi_1$, **I** leads **V** by ϕ (Figure 8.19(a))

When the resolved component of $\mathbf{I_{RL}}$ in the vertical direction has a magnitude greater than that of $\mathbf{I_C}$:

$I_C < I_{RL} \sin \phi_1$, **I** lags **V** by ϕ (Figure 8.19(b))

When the resolved component of $\mathbf{I_{RL}}$ in the vertical direction has a magnitude equal to that of $\mathbf{I_C}$:

$I_C = I_{RL} \sin \phi_1$, **I** and **V** are in phase (Figure 8.19(c))

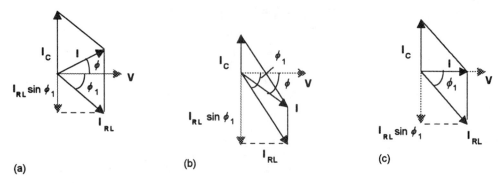

Figure 8.19 *The three possible conditions*

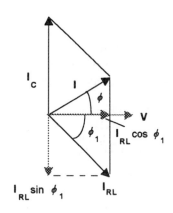

Figure 8.20 *Resolving
the current through RL*

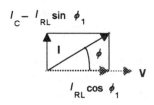

Figure 8.21 *Determining
the circuit current*

We can determine the total circuit current **I** by a scale drawing of the phasors to give one of the phasor diagrams in Figure 8.19. Alternatively, we can resolve the current phasor **I**$_{RL}$ into horizontal and vertical component phasors: **I**$_{RL}$ cos ϕ_1 horizontally and vertically **I**$_{RL}$ sin ϕ_1. The horizontal component is in phase with **V** and is termed the 'in phase' component, the other component is at 90° and is termed the 'quadrature' component. For Figure 8.19(a) we thus have the situation shown in Figure 8.20. The vertical current has the magnitude of $I_C - I_{RL}$ sin ϕ_1 and the horizontal current is I_{RL} cos ϕ_1. We can thus draw the phasor diagram shown in Figure 8.21 and so:

$$I^2 = (I_{RL} \cos \phi_1)^2 + (I_C - I_{RL} \sin \phi_1)^2 \qquad [46]$$

$$\tan \phi = \frac{I_C - I_{RL} \sin \phi_1}{I_{RL} \cos \phi_1} \qquad [47]$$

$$\cos \phi = \frac{I_{RL} \cos \phi_1}{I} \qquad [48]$$

Example

A circuit consists of a coil with inductance 150 mH and resistance 40 Ω in parallel with a capacitor of capacitance 30 µF. A 240 V r.m.s., 50 Hz voltage supply is connected across the circuit. Determine the current in (a) the coil, (b) the capacitor and (c) drawn from the supply.

(a) The inductive reactance is $X_L = 2\pi f L = 2\pi \times 50 \times 0.150 = 47.1$ Ω. The impedance Z_{RL} of the coil has thus the value:

$$Z_{RL} = \sqrt{R^2 + X_L^2} = \sqrt{40^2 + 47.1^2} = 61.8 \text{ Ω}$$

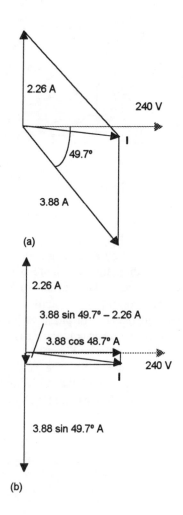

(a)

(b)

Figure 8.22   *Example*

The current through the coil has the magnitude:

$$I_{RL} = \frac{V}{Z_{RL}} = \frac{240}{61.8} = 3.88 \text{ A}$$

with the phase angle:

$$\phi_1 = \tan^{-1}\frac{X_L}{R} = \frac{47.1}{40} = 49.7°$$

This is the angle by which the current lags the supply voltage.
(b) The capacitive reactance is $X_C = 1/2\pi f C = 1/(2\pi \times 50 \times 30 \times 10^{-6}) = 106.1 \ \Omega$. The current through the capacitor has thus the magnitude:

$$I_C = \frac{V}{X_C} = \frac{240}{106.1} = 2.26 \text{ A}$$

The current leads the supply voltage by 90°.
(c) Figure 8.22(a) shows the phasor diagram for the currents in the circuit and Figure 8.22(b) the phasor diagram when I_{RL} has been resolved to the in-phase and quadrature components. Hence:

$$I = \sqrt{(3.88 \sin 49.7° - 2.26)^2 + (3.88 \cos 48.7°)^2} = 2.65 \text{ A}$$

Its phase angle by which it lags the supply voltage is:

$$\phi = \tan^{-1}\frac{3.88 \sin 49.7° - 2.26}{3.88 \cos 48.7°} = 15.2°$$

Revision

13 A resistance of 100 Ω is in parallel with an inductance of 100 mH. What will be the current drawn from the supply and the impedance of the circuit when a voltage of 12 V r.m.s., 50 Hz is applied to it?

14 A resistance of 1 kΩ is in parallel with a capacitance of 2 μF. What will be the current drawn from the supply and the impedance of the circuit when a voltage of 12 V r.m.s., 50 Hz is applied to it?

15 A coil with a resistance of 10 Ω and inductance 100 mH is in parallel with a capacitor. What value of capacitance will be needed if the overall circuit current and voltage are to be in phase when the supply voltage has a frequency of 50 Hz?

8.7 Power The power developed in a d.c. circuit is the product of the current and the resistance. With an a.c. circuit the voltage and current are varying with time, thus we refer to the instantaneous power p as the product of the instantaneous current i and instantaneous voltage v:

$$p = iv \qquad\qquad [49]$$

Since the current and voltage are changing with time, the product varies. For this reason we quote the average power dissipated over a cycle.

8.7.1 Power for a purely resistive circuit

For a purely resistive circuit the current and voltage are in phase and we have $i = I_m \sin \omega t$ and $v = V_m \sin \omega t$, where i and v are the instantaneous values and I_m and V_m the maximum values. Hence the instantaneous power p is:

$$p = iv = I_m \sin \omega t \times V_m \sin \omega t = I_m V_m \sin^2 \omega t \qquad [50]$$

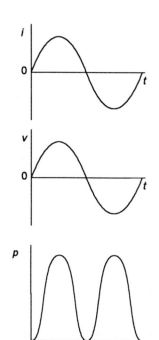

Figure 8.23 shows the graphs of i, v and p as functions of time t. By inspection of the graph of p varying with time we might deduce that, since the graph is symmetrical about a horizontal axis through its middle, the average value of p over one cycle is:

$$\text{average power } P = \tfrac{1}{2}I_m V_m \qquad [51]$$

Alternatively, we can use the trigonometric relation $\cos 2\theta = 1 - 2\sin^2 \theta$ to write equation [50] as:

$$p = I_m V_m \tfrac{1}{2}(1 - \cos 2\omega t) = \tfrac{1}{2}I_m V_m - \tfrac{1}{2}I_m V_m \cos 2\omega t \qquad [52]$$

The average value of a cosine function over one cycle is zero. Hence:

$$\text{average power } P = \tfrac{1}{2}I_m V_m \qquad [53]$$

Figure 8.23   *Current, voltage and power as functions of time*

The r.m.s. current $I_{r.m.s.}$ is $I_m/\sqrt{2}$ and the r.m.s. voltage $V_{r.m.s.}$ is $V_m/\sqrt{2}$. Hence we can write the equation for the average power dissipated with a purely resistive circuit as:

$$\text{average power } P = I_{r.m.s.}V_{r.m.s.} \qquad [54]$$

Example

Determine the average power developed when a 240 V r.m.s. alternating current supply is connected across a resistance of 100 Ω.

Since $V_{r.m.s.} = RI_{r.m.s.}$ then equation [54] can be written as:

$$\text{average power} = I_{r.m.s.}V_{r.m.s.} = \frac{V_{r.m.s.}^2}{R} = \frac{240^2}{100} = 576 \text{ W}$$

Revision

16   Calculate the resistance of an electric light bulb that develops a power of 100 W when connected across a 240 V alternating current supply.

8.7.2 Power for a purely reactive circuit

Consider a circuit which is purely reactive. For such a component there is a phase angle of 90° between the current and the voltage (Figure 8.24). For example, we might have $i = I_m \sin \omega t$ and $v = V_m \sin (\omega t - 90°)$. The instantaneous power p is the product of the instantaneous current and voltage:

$$p = I_m \sin \omega t \times V_m \sin (\omega t - 90°) \qquad [55]$$

Figure 8.24 shows i, v and p as functions of time. Over one cycle the average power for the purely reactive element is zero.

We can demonstrate that this is the case by using the trigonometric relation $2 \sin A \sin B = \cos (A - B) - \cos (A + B)$, this giving:

$$2 \sin \omega t \times \sin (\omega t - 90°) = \cos 90° - \cos (2\omega t - 90°)$$

$$= 0 - \sin 2\omega t$$

and hence:

$$p = \tfrac{1}{2} I_m V_m \sin 2\omega t \qquad [56]$$

The average value of $\sin 2\omega t$ over a cycle is 0 and thus the average power is 0.

Average power = 0   [57]

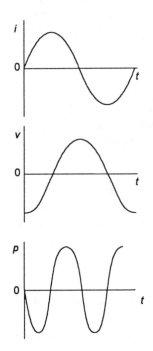

Figure 8.24   *Current, voltage and power as functions of time*

8.7.3 Power for a circuit having resistance and reactance

For a component or circuit having both resistance and reactance, there is a phase difference other than 90° between the current and the voltage. thus we might have $i = I_m \sin \omega t$ and $v = V_m \sin (\omega t - \phi)$ with ϕ being the phase difference (Figure 8.25). If we resolve the voltage phasor into horizontal and vertical components of $V \cos \phi$ and $V \sin \phi$, then the component at 90° to **I** is the situation we would get with a purely reactive circuit and so the average power over a cycle is zero for that component. The component in phase with the current is the situation we get with a purely resistive circuit and so the average power is given by using equation [54]. Thus the average power P over a cycle is given by the in-phase components as:

$$P = IV \cos \phi \qquad [58]$$

We can obtain the above relationship by using trigonometric relations. Thus for:

$$p = I_m \sin \omega t \times V_m \sin (\omega t - \phi) \qquad [59]$$

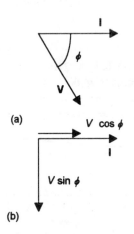

Figure 8.25   *Phasors*

we can use the relation $2 \sin A \sin B = \cos (A - B) - \cos (A + B)$ and so $2 \sin \omega t \sin (\omega t - \phi) = \cos \phi - \cos (2\omega t - \phi)$. Hence:

$$p = \tfrac{1}{2} I_m V_m \cos \phi - \tfrac{1}{2} I_m V_m \cos (2\omega t - \phi) \qquad [60]$$

The average value of $\cos (2\omega t - \phi)$ over one cycle is zero and thus:

$$\text{average power } P = \tfrac{1}{2} I_m V_m \cos \phi = I_{r.m.s.} V_{r.m.s.} \cos \phi \qquad [61]$$

where I and V are r.m.s. values, the average power being in watts. The term *power factor* is used for $\cos \phi$, since it is the factor by which the power for a purely resistive circuit is multiplied to give the power when there is both resistance and reactance. See Section 8.7.4 for further discussion of the power factor.

Example

Determine the power dissipated when a coil of inductance 40 mH and resistance 20 Ω is connected to a 240 V r.m.s., 50 Hz supply.

The inductive reactance is $X_L = 2\pi f L = 2\pi \times 50 \times 0.040 = 12.6 \; \Omega$. Thus the impedance $Z = \sqrt{(20^2 + 12.6^2)} = 23.6 \; \Omega$ and so the current I = 240/23.6 = 10.2 A. For a series RL circuit we have $\cos \phi = R/Z$ = 20/23.6 = 0.85. Thus the power factor is 0.85. The average power dissipated is thus $IV \cos \phi = 10.2 \times 240 \times 0.85 = 2081$ W. Alternatively we could recognise that power will only be dissipated in the resistance, no power being dissipated in the reactance. Thus the power developed is $I^2R = 10.2^2 \times 20 = 2081$ W.

Example

Determine the resistance and inductance of a series RL circuit if, when connected to a 240 V r.m.s., 50 Hz supply, it takes a power of 0.8 kW and a current of 4 A?

The circuit has a resistive component of R and a reactive component of X_L. The power is the power dissipated in just the resistive component and thus $P = I^2R$ and so $R = 800/4^2 = 50 \; \Omega$. The circuit impedance $Z = V/I = 240/4 = 60 \; \Omega$. Since $Z = \sqrt{(R^2 + X_L^2)}$ then $X_L = \sqrt{(Z^2 - R^2)} = \sqrt{(60^2 - 50^2)} = 33.2 \; \Omega$. Since the reactance $X_L = 2\pi f L$ then $L = 33.2/(2\pi \times 50) = 0.11$ H.

Revision

17 Determine the power dissipated when a coil of inductance 100 mH and resistance 60 Ω is connected across a 12 V r.m.s., 100 Hz supply.

18   Determine the power dissipated with an 8 μF capacitor in series with a 1 kΩ resistor when the series arrangement is connected across a 12 V r.m.s., 50 Hz supply.

8.7.4 Apparent power, power factor and reactive power

The average *active power*, the term *true power* is often used, dissipated in an alternating current circuit is given by equation [61] as:

$$\text{average active power } P = I_{\text{r.m.s.}} V_{\text{r.m.s.}} \cos \phi$$

The product of the r.m.s. values of the current and voltage is called the *apparent power S*, the unit of apparent power being volt amperes (V A). Multiplying the apparent power by $\cos \phi$ gives the true power dissipated. For this reason, $\cos \phi$ is called the *power factor*. It has no units. Thus:

$$\text{average power} = \text{apparent power} \times \text{power factor} \qquad [62]$$

For a purely reactive circuit the power factor is 0, for a purely resistive factor it is 1. A circuit in which the current lags the voltage, i.e. an inductive circuit, is said to have a lagging power factor; a circuit in which the current leads the voltage, i.e. a capacitive circuit, is said to have a leading power factor.

If we consider an alternating current circuit such as a *RL* series circuit, the overall voltage will be leading the current by some phase angle ϕ. We can resolve the voltage phasor into two components, one in phase with the current of magnitude $V \cos \phi$ and one, at a phase of 90° to the current, of magnitude $V \sin \phi$ (Figure 8.26(a)). Multiplying the magnitude of each of these by the magnitude of the current I transforms the phasor diagram into a power triangle (Figure 8.26(b)). For this triangle, the horizontal is the active power $P = IV \cos \phi$, the hypotenuse is the apparent power $S = IV$ and the vertical is $VI \sin \phi$, this being termed the *reactive power Q* and given the unit of volt amperes reactive (V Ar).

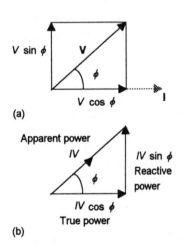

Figure 8.26   *(a) Voltage components, (b) power diagram*

Example

A load takes 120 kW at a power factor of 0.6 lagging. Determine the apparent power and the reactive power.

The true power $P = 120$ kW. Thus $120 = VI \cos \phi = VI \times 0.6$ and so the apparent power $S = IV = 120/0.6 = 200$ kV A.

The reactive power $Q = IV \sin \phi$. Since $\cos \phi = 0.6$ then $\phi = 53.1°$ and so $\sin \phi = 0.8$ and $Q = 200 \times 0.8 = 160$ kV Ar.

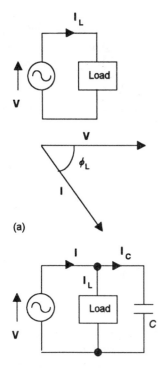

(a)

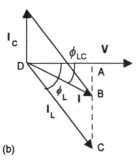

(b)

Figure 8.27 *Improving the power factor*

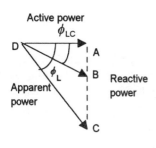

Figure 8.28 *Power triangles*

Revision

19 A coil with an inductance of 1 H and a resistance of 200 Ω is connected to a 240 V r.m.s. 50 Hz supply. Determine the active power and the apparent power.

20 A coil with an inductive reactance of 10 Ω and a resistance of 7.5 Ω is connected to a 240 V r.m.s., 50 Hz supply. Determine the power factor, the apparent power and the active power.

8.7.5 Power factor improvement

The active power is the power content of the apparent power that is delivered by a supply system. The lower the power factor the greater is the current needed to produce the same active power. The higher the current, the greater the size of cable required to deliver the supply. Thus it is more economical in the delivery of power to have a system with a large power factor. Since most residential and industrial loads are inductive and have a lagging power factor, power factors can be improved by the connection of capacitors in parallel with the load.

Figure 8.27(a) shows the situation with an inductive load and Figure 8.27(b) the same load when a capacitor is connected in parallel with it. With no capacitor the current supplied I is the same as the current through the load I_L and we have a lagging phase angle of ϕ_L and thus a lagging power factor of $\cos \phi_L$. When the capacitor is connected in parallel, the current supplied gives a current through the load and a current through the capacitor. I is thus the phasor sum of I_C and I_L. The current I taken from the supply is now at a lower lagging phase angle ϕ_{LC} and hence a higher power factor.

Initially, Figure 8.27(b) gives $AC = I_L \sin \phi_L$. After connecting the capacitor, $AB = I \sin \phi_{LC}$. Since $BC = I_C$ then:

$$I_C = AC - AB = I_L \sin \phi_L - I \sin \phi_{LC} \qquad [63]$$

Multiplying this equation by V gives:

$$VI_C = VI_L \sin \phi_L - VI \sin \phi_{LC} \qquad [64]$$

But $VI_L \sin \phi_L$ is the reactive power with no capacitor Q_L and $VI \sin \phi_{LC}$ is the reactive power Q_{LC} with the capacitor. Thus:

$$VI_C = Q_L - Q_{LC} = \text{change in reactive power} \qquad [65]$$

The triangles ABD and ACD with their sides multiplied by V give the power triangles (Figure 8.28). AB is the reactive power with the capacitor, AD is the reactive power without the capacitor, DC is the apparent power without the capacitor, DB is the apparent power with the capacitor and DA, since it is the apparent power multiplied by the cosine of the phase angle, which is the active power.

Example

A 240 V r.m.s., 50 Hz a.c. motor gives an active power output of 1 kW when operating at a power factor of 0.8 lagging. Calculate the reduction in the current taken by the motor if the power factor is improved to 0.9 lagging.

At a power factor of 0.8 lagging, the current taken by the motor from the supply is given by equation [6] as:

$$I_{r.m.s.} = \frac{P}{V_{r.m.s.} \cos\phi} = \frac{1 \times 10^3}{240 \times 0.8} = 5.21 \text{ A}$$

At a power factor of 0.9 lagging, the current taken is:

$$I_{r.m.s.} = \frac{P}{V_{r.m.s.} \cos\phi} = \frac{1 \times 10^3}{240 \times 0.9} = 4.62 \text{ A}$$

The current is thus reduced by 5.21 − 4.62 = 0.59 A.

Example

Calculate the capacitance required for a capacitor which is to be connected in parallel with a load of 500 W, supplied by a power supply of 240 V, 50 Hz, to change the power factor from 0.75 to 0.85 lagging.

The apparent power with the power factor 0.75 is 500/0.75 = 666.7 V A. With the power factor of 0.75 we have $\cos\phi$ = 0.75 and so ϕ = 41.4°. Hence the reactive power is 666.7 × sin 41.4° = 440.9 V Ar. With the power factor of 0.85, the apparent power is 500/0.85 = 588.2 V A, ϕ = 31.8° and the reactive power is 588.2 sin 31.8° = 310.0 V Ar. The change in reactive power required is thus 440.9 − 310.0 = 130.9 V Ar. Using equation [65]:

$$VI_C = \text{change in reactive power} = 130.9 \text{ V Ar}$$

The current taken by the capacitor is thus 130.9/240 = 0.545 A. Since current is the voltage across the capacitor divided by its reactance, the reactance is 240/0.545 = 440.4 Ω. Thus 440.4 = $1/2\pi f C$ and so C = 7.23 μF.

Example

A load takes a current of 4 A r.m.s. when connected to a 240 V r.m.s., 50 Hz supply and has a power factor of 0.75 lagging. What capacitance will a capacitor need to have if, when connected in parallel with the load, it is to increase the power factor to 0.85 lagging?

Since the initial power factor $\cos\phi_L$ is 0.75, the initial phase angle is ϕ_L = 41.4°. The corrected power factor is 0.85 and so the corrected

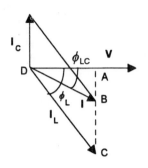

Figure 8.29 *Example*

phase angle is $\phi_{LC} = 31.8°$. The phasor diagram for the currents is shown in Figure 8.29. For triangle ACD we have $AC = I_L \sin \phi_L = 4 \times \sin 41.4° = 2.65$ A. For triangle ABD we have AB = AD tan ϕ_{LC} and since AD = $I_L \cos \phi_L = 4 \times 0.75 = 3$ then AB = 3 tan 31.8° = 1.86 A. Thus I_C = AC – AB = 2.65 – 1.86 = 0.79 A. For the capacitor we have $X_C = V/I_C = 240/0.79 = 304$ Ω and so $1/2\pi f C = 304$ and $C = 10.5$ μF.

Revision

21 Calculate the reactive power required of a capacitor which when connected in parallel with a load of 650 kV A will change its power factor from 0.7 to 0.9 lagging.

22 A load takes a current of 6 A r.m.s. when connected to a 240 V r.m.s., 50 Hz supply and has a power factor of 0.65 lagging. What capacitance will a capacitor need to have if, when connected in parallel with the load, it is to increase the power factor to 0.90 lagging?

8.8 Circuit analysis using PSpice

SPICE (*S*imulation *P*rogram with *I*ntegrated *C*ircuit *E*mphasis) is a computer program that has been used since the 1970s for the analysis of circuits. PSpice is the version developed for the personal computer. Two versions are currently in use, a DOS version and a Windows version. With the Windows version circuits are described by drawing a labelled circuit on the screen, with the DOS version the circuits are described by writing a list describing the components and their connections. The following is a brief indication of how PSpice might be used for the circuit analysis.

For the Windows version:

1 Start the PSpice drawing program by clicking on the Schematics icon in the Design Center Eval 6.0 window.

2 The screen display then shows the top line of:

Schematics: <new> p.1   Part:

After the drawing is named, the name replaces <name>. p.1 indicates that it is page 1 of the drawing. Part: gives the reference of the last part chosen from the library of parts. Underneath it are the headings for pull-down menus, the menus being dropped down when a header is clicked on from the header list of:

File Edit Draw Navigate View Options Analysis Tools Markers Help=F1

3 Click Draw and then, on the resulting drop down menu, Get Part. A pop-up window for Add Part appears on the screen. If the part name

is known it can be typed in the Part box; for a resistor this is R. Click on OK and a cursor controlled resistor symbol appears in the drawing area. If the part name is not known, click Browse. This results in a Get Part pop up window appearing on the screen. The right-hand part of the window shows the libraries of parts available. For the parts required for circuits in this chapter, click on analog.slb. Then click on the down arrow of the Part box until the component scrolls into view. Click on it and the R symbol then appears in the Part name: box. Click on OK and a cursor controlled resistor symbol appears in the drawing area.

4   To place the resistor in the required place on the screen, move the mouse and when it is in the correct position click on the left mouse button. The cursor still is in the form of a resistor symbol and a further resistor can be placed on the screen by moving the mouse and clicking the left mouse button when in the appropriate place. This can be continued until the right mouse button is pressed. A resistor can be rotated by selecting it, as a result of pointing to it with the mouse and clicking, holding down the <Ctrl> key and pressing r.

The default value for a resistor is 1 kΩ and the default name is R(number). The value can be changed by double clicking it. This results in a window opening to allow a new value to be entered. This can be done by typing the value out in full, e.g. 1000, or using exponential notation, e.g. 10^3 which is entered as 1e3, or using prefixes such as U for micro, M for milli, K for kilo, MEG for mega, e.g. 1000 is 1K. Units are not entered since resistance is assumed to be in ohms.

5   Voltage or current sources can be selected by getting the part and positioning it with the mouse. For example, if the part is not known the sequence is Draw, Get New Part, Browse, look in the library and select source.slb. For a d.c. voltage source select the part VSRC, select OK. For a sinusoidal voltage waveform select the part VSIN, select OK. To set the values for a source such as the sine wave generator, double clock on the supply symbol. This opens a window in which values can be entered. Values are required for voff, vampl and freq. Voff is the d.c. offset voltage for the a.c. supply; for a.c. that alternates about the zero voltage axis set this value to 0 by double clicking the voff line in the window and then type the value in the value box. Vampl is the amplitude, i.e. the maximum value, of the a.c. signal. Double click the vampl line in the window and then type the value in the value box. If you want 10 V, do not type the 10 with a space between it and the V but type as 10V. Freq is the frequency of the a.c. signal. double click the freq line in the window and then type the value in the value box. Three other optional values can be set: time delay td for the delay in seconds between the time when the examination of the circuit begins and the source is turned on, usually set to 0; damping factor df for the exponential decay of the signal, set to 0 for the work in this chapter; phase is used to set

the phase, in degrees, and is usually set to 0 for a voltage source which is used as the reference. These three values can be set or left unused.

6   To draw connecting wires select Draw and then Wire. This changes the cursor to a pencil. Move the pencil to the start point of the circuit and click, move to any bend point and click, move to the end point of the wire, e.g. at a resistor, and click. Click the right mouse button to stop the wire. Double click the right mouse button to start the wiring operation for further wires in the circuit.

7   Devices should be placed in the circuit at the points where the voltage and current are required. To monitor the voltage select from the library the VIEWPOINT and position at a node on the circuit. For current, select IPROBE from the library and wire it into the circuit like an ammeter. PSpice will give nodes numbers but if you want simpler numbers, double click a wire at a node and the resulting pop-up window allows a number to be entered.

8   The schematic can be named and saved by selecting File, Save As and then typing a suitable name in the File Name box.

9   To carry out the analysis of an a.c. circuit and display the wave-forms: select Analysis from the top of the screen and then Setup from the drop-down menu. From the window that opens select Transient by clicking on its enable square. This opens another window. The Print Step value is used to set the time interval between values that are to be printed in the output or displayed graphically on the screen. the Final time is the length of time for which the analysis will continue. Set the values and then click OK.

     Since there is data to be graphed select Analysis and then Probe Setup. Probe is the graphical program. Click on the enabling circle for Run probe. Then click OK. The Probe window shows a blank graph. Clicking the sequence Trace\Add\V(1)\OK adds the source voltage to the plot. Similarly Trace\Add\ then typing V(1,2) for the voltage between nodes 1 and 2\OK can be used to display the voltage occurring between specific nodes in the circuit.

10   If we want current and voltage values sent to an output file, i.e. a printout is required of the voltage and current values, we select from the library special devices called VPRINT (Figure 8.30). VPRINT1 has one connector and is connected to a node to give the maximum voltage of a.c. relative to earth, VPRINT2 has two connectors and is connected across a component, like a voltmeter would be, to give the maximum a.c. voltage drop for that component. IPRINT is selected and inserted into the circuit as an ammeter would be to give a reading of the maximum a.c. current at that point. Each of these devices must be configured to output a.c. values. Double click on a device to open a window and then select the relevant attributes. The analysis is then carried out as by selecting Analysis and then Setup.

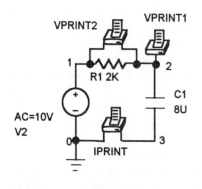

Figure 8.30   *Output devices*

AC Sweep is then enabled by clicking in its box and AC Sweep set to linear and the Sweep parameters to Total Pts. 1, and if the analysis is to be carried out at a particular frequency the Start Freq. and End Freq. values are set to this value. Then select OK. To run the analysis select Analysis and then Simulate; the results appear in a file with the same root name but with an .OUT extension.

For the DOS version:

1   Initially a circuit drawing should be made with each node in the circuit labelled. Node 0 is taken as the reference node for the circuit and all voltages at other nodes are relative to that node. Each resistor should be labelled with R followed by some number or other letters, e.g. R1, R2, R3, etc. Voltage sources should be labelled as V followed by some number or other letters, e.g. V1, V2, etc. Similarly each capacitor should be labelled with C and each inductor by L.

2   Open a word processor program or a text editor program.

3   The input file for PSpice must include a format statement for each circuit component and each voltage source. For a resistor the format statement has the form:

R[name] [no. of first node] [no. of second node] [resistance]

Thus for a resistance R2 of 100 Ω to be connected between nodes 1 and 2 the statement would be:

R2 1 2 100 OHMS

For a voltage source the format statement has the form:

V[name] [no. of first node] [no. of second node]
        <type of source> [VO   VA   Freq.   TD   Theta]

The first node is taken to be the positive node, the second node the negative node. For a sinusoidal voltage source V1 with no offset value VO, i.e. the signal alternates about the zero voltage axis, an a.c. voltage of maximum value VA = 5 V between nodes 0 and 1, a frequency of 1000 Hz, no time delay TD, i.e. the signals start at time $t = 0$ and with the applied sine wave retaining a constant amplitude, theta then being set to 1, the statement is:

V1   1   0   SIN (0   5   1000   0   0)

Write a program listing each of the format statements needed to define the circuit. A comment line, preceded by *, can be used to describe the program for the benefit of the user. Such comment lines are not used in the operation of the program.

4   A statement needs to be added to the program in order to indicate the type of analysis to be carried out. For an analysis which plots current and voltage graphs a transient analysis is required to show the variation with time. Its format is:

.TRAN[time interval] [final time] UIC

It defines the time interval between data points and determines the final time of the analysis. Thus for two cycles of a 1000 Hz waveform sampled at 50 data points, i.e. since the period is 0.001 s then time interval is $0.001/25 = 40 \times 10^{-6}$ s or 40E–3 and the final time is $0.002 = 2 \times 10^{-3}$ s or 2E–3, the statement would be:

.TRAN 40E–6 2E–3 UIC

5   The data to be outputted from the analysis, e.g. all the node voltage values relative to the 0 node or the voltages between nodes, is requested by lines in the program. Such statements have the form:

.PRINT [Analysis type] [output] [output] [output]

Thus we might have:

.PRINT TRAN V(1) V(2) V(3)

to give the voltages at particular nodes and:

.PRINT TRAN V(1,2) V(2,3)

to give the voltages between nodes with:

.PRINT TRAN I(R1)

to give the current through resistor R1.

6   The end of the program must be indicated by the .END statement. An example of a program to plot graphs of voltage and current for an a.c. circuit involving a series resistance and capacitance (Figure 8.31) is:

```
FIG8.31: SERIES RC CIRCUIT
VSOURCE 1 0 SIN(0 10 1000 0 0)
C1 1 2 100UF  IC = 0
R2 2 0 5
*CONTROL STATEMENT FOLLOWS
.TRAN 40E–6 2R–3 UIC
.PRINT TRAN V(1) V(1, 2) V(2) I(C1)
.PROBE
.OPTIONS NOPAGE
.END
```

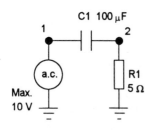

Figure 8.31   *RC circuit*

The IC = 0 on the end of the capacitor statement is to indicate that the initial value of the voltage across it is 0. The .PROBE statement calls up the computer graphics program in Pspice to enable the waveforms to be seen on the screen. The .OPTIONS NOPAGE statement is to ensure that only the first part of the output file contains the header.

7   Save the input data file, giving it a name.

8   Open the PSpice program. There will be a prompt asking you to enter the name of the input data PSpice file. The result will be an output file with the same name as the input file but with the extension changed to .OUT.

Problems   1   Determine the capacitive reactance of:
(a) a capacitor of capacitance 0.1 μF when the frequency of the current through it is 1 kHz,
(b) a capacitor of capacitance 100 μF when the voltage across it is 12 sin 5000 t V,
(c) a capacitor of capacitance 20 pF when the current through it is 100 sin 12000t mA.

2   Determine the inductive reactance of:
(a) an inductor of inductance 20 mH when the frequency of the current through it is 20 kHz,
(b) an inductor of inductance 1 H when the voltage across it is 10 sin 4000t V,
(c) an inductor of inductance 100 mH when the current through it is 20 sin 100t mA.

3   Determine the current through a pure capacitor of capacitance 0.01 μF when the voltage across it is 40 sin 1000t V.

4   At what frequency will the inductive reactance of an inductor of inductance 15 mH be 2.5 kΩ?

5   At what frequency will the capacitance reactance of a capacitor of capacitance 20 μF be 100 Ω?

6   A coil of inductance 50 mH and resistance 15 Ω is connected across a 240 V r.m.s., 50 Hz supply. Determine the circuit impedance, the current taken and the phase angle between the current and voltage.

7   A coil has a resistance of 15 Ω and an inductive reactance of 8 Ω. Determine the impedance of the coil and the phase angle between the voltage across the coil and the current through it.

8   A voltage of 100 V r.m.s., 50 Hz is connected to a series RL circuit, the resistance being 45 Ω and the inductance having an inductive

reactance of 60 Ω. Determine (a) the circuit impedance, (b) the current through the circuit, (c) the voltages across each component.

9   A voltage of 100 V r.m.s., 50 Hz is connected to a series RL circuit, the resistance being 15 Ω and the inductance being 60 mH. Determine (a) the circuit impedance, (b) the magnitude and phase angle of the current through the circuit.

10   A voltage of 50 V r.m.s., 50 Hz is applied to a series RC circuit. If the current taken is 1.7 A when the voltage drop across the resistor is 34 V, what is the capacitance of the capacitor?

11   A voltage of 240 V r.m.s., 50 Hz is applied to a series RC circuit. If the resistance is 40 Ω and the capacitance 50 μF, determine (a) the circuit impedance, (b) the magnitude and phase angle of the current.

12   A voltage of 24 V r.m.s., 400 Hz is applied to a series RC circuit. If the resistance is 40 Ω and the capacitance 15 μF, determine (a) the circuit impedance, (b) the magnitude and phase angle of the current.

13   A series RLC circuit takes a current of 5 A r.m.s. If the resistance is 10 Ω, the inductive reactance 20 Ω and the capacitive reactance 10 Ω, what will be (a) the magnitude and phase of the voltage applied to the circuit, (b) the voltages across each component?

14   A RLC circuit has a coil having a resistance of 75 Ω and an inductance of 0.15 H in series with a capacitance of 8 μF. If a voltage of 100 V r.m.s., 200 Hz is applied to it, determine (a) the magnitude and phase angle of the current, (b) the voltage across the coil, (c) the voltage across the capacitor.

15   A circuit consists of a resistance of 10 Ω in parallel with an inductance of 50 mH. What is the magnitude and phase angle of the current taken from the supply when a voltage of 240 V r.m.s., 50 Hz is applied?

16   A circuit consists of a resistance of 40 Ω in parallel with an inductance of reactance 30 Ω. What is the magnitude and phase angle of the current taken from the supply when a voltage of 120 V r.m.s., 50 Hz is applied?

17   A circuit consists of a resistance in parallel with capacitance. What are the values of the resistance and capacitance if, when a voltage of 240 V r.m.s., 200 Hz is applied, the current taken from the supply is 2 A with a phase angle of 53.1° leading?

18   A resistance is connected in parallel with a capacitance. If, when a voltage of 100 V r.m.s. is applied to the circuit, the current drawn from the supply is 2 A at a phase angle of 30° leading, what are (a) the magnitudes of the currents through the resistance and the

capacitance, and (b) the values of the resistance and the capacitive reactance?

19 Determine the power dissipated when a coil of inductance 100 mH and resistance 40 Ω is connected across a 10 V, 500 Hz supply.

20 A coil has an inductance of 10 H and resistance of 250 Ω and is connected in series with a 5 μF capacitor across a 240 V, 50 Hz supply. Determine the power factor, the apparent power, the reactive power and the active power.

21 A circuit consists of a coil, resistance 200 Ω and inductance 10 H, in parallel with a capacitance of 5 μF. A voltage of 240 V r.m.s., 50 Hz is applied to the circuit. Determine (a) the current drawn from the supply, (b) the power factor of the circuit, (c) the active power, (d) the apparent power, (e) the reactive power.

22 A series *RC* circuit has a resistance of 2 kΩ and a capacitance of 1 μF and a 240 V r.m.s., 50 Hz supply is applied to the circuit. Determine (a) the power factor, (b) the active power, (c) the apparent power, (d) the reactive power.

23 A load takes 80 kW at a power factor of 0.6 lagging. Determine the apparent power and the reactive power.

24 Calculate the reactive power required of a capacitor which when connected in parallel with a load of 300 kV A will change its power factor from 0.7 to 0.9 lagging.

25 A load takes a current of 5 A r.m.s. when connected to a 240 V r.m.s., 50 Hz supply and has a power factor of 0.7 lagging. What capacitance will a capacitor need to have if, when connected in parallel with the load, it is to increase the power factor to 0.9?

9 Complex numbers

9.1 Introduction

In Chapter 8 simple a.c. circuits were analysed by the use of phasor diagrams. This method, however, becomes rather complicated for all but simple circuits. The analysis of more complex circuits is considerably simplified by the use of complex numbers and this chapter is an introduction to their use for this purpose. The analysis is preceded by a discussion of complex numbers and their manipulation.

9.2 Complex numbers

The following is a discussion of the terms imaginary number and complex number and the basic arithmetic involved in the use of complex numbers.

9.2.1 Imaginary numbers

If we square the number +2 we obtain +4, if we square the number −2 we obtain +4. Thus, reversing these operations, the square root of +4 is either +2 or −2. But what is the square root of −4? To give an answer we need another form of number. If we invent a special number of $j = \sqrt{-1}$ (mathematicians often use i rather than j but engineers and scientists generally use j to avoid confusion with i used for current in electrical circuits), then we can write $\sqrt{-4} = \sqrt{-1} \times \sqrt{4} = \pm j2$.

> A number which is a multiple of j is termed *imaginary*.

$j = \sqrt{-1}$ and thus $j^2 = -1$. Since j^3 can be written as $j^2 \times j$ then $j^3 = -j$. Since j^4 can be written as $j^2 \times j^2$ then $j^4 = +1$.

$$j = \sqrt{-1}, \quad j^2 = -1, \quad j^3 = -j, \quad j^4 = +1 \tag{1}$$

9.2.2 Complex numbers

The solution of a quadratic equation of the form $ax^2 + bx + c = 0$ is given by:

$$x = \frac{-b \pm \sqrt{b^2 - 4ac}}{2a} \tag{2}$$

Thus if we want to solve the quadratic equation $x^2 - 4x + 13 = 0$ then:

$$x = \frac{4 \pm \sqrt{16 - 52}}{2} = 2 \pm \sqrt{-9}$$

We can represent $\sqrt{-9}$ as $\sqrt{-1} \times \sqrt{+9} = j3$. Thus the solution can be written as $2 \pm j3$, a combination of a real and plus or minus an imaginary number.

> The term complex number is used for a combination by addition or subtraction of a real number and a purely imaginary number.

In general a complex number z can be written as:

> $$z = a + jb \qquad\qquad [3]$$

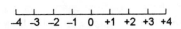

Figure 9.1   *The real number line*

where a is the real part of the complex number and b the imaginary part.

9.2.3 The Argand diagram

Real numbers can be represented as points along a line in the way shown in Figure 9.1; negative numbers extend to the left of the zero point and positive numbers to the right. The effect of multiplying a real number by (-1) is to move the point on the number line from one side of the origin to the other. Figure 9.2 illustrates this for $(+2)$ being multiplied by (-1). We can think of the positive number line radiating out from the origin being rotated through 180° to its new position after being multiplied by (-1). But $(-1) = j^2$. Thus multiplication by j^2 is equivalent to a 180° rotation. Multiplication by j^4 is a multiplication by $(+1)$ and so is equivalent to a rotation through 360° (Figure 9.3). On this basis it seems reasonable to take a multiplication by j to be equivalent to a rotation through 90° (Figure 9.4) and a multiplication by j^3 a rotation through 270°. This concept of multiplication by j as involving a rotation is the basis of the use of complex numbers to represent phasors in alternating current circuits.

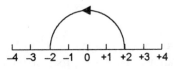

Figure 9.2 $(+2) \times (-1)$

> Multiplication by j is a rotation through 90°.
> Multiplication by j^2 is a rotation through 180°.
> Multiplication by j^3 is a rotation through 270°.
> Multiplication by j^4 is a rotation through 360°.

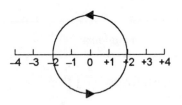

Figure 9.3 $(+2) \times (+1)$

The above discussion leads to a way of representing complex numbers. The real part of the complex number is represented by the distance to the right or left of the vertical line through the zero point while the imaginary part of the complex number is represented by a distance upwards or downwards from the horizontal line through the zero point (Figure 9.5), i.e. the complex number is represented by its rectangular (Cartesian) co-ordinates on the diagram. Such a diagram is called an

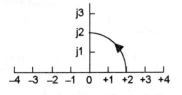

Figure 9.4 $(+2) \times j$

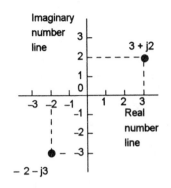

Figure 9.5   *Argand diagram*

Argand diagram. Figure 9.3 shows how the complex numbers 3 + j2 and −2 − j3 can be represented on such a diagram. Thus 3 + j2 is a point +3 units from the vertical line through the zero point and +2 units from the horizontal line through the zero point. −2 − j3 is a point −2 units from the vertical line through the zero point and −3 units from the horizontal line through the zero point.

> The Argand diagram is used to represent pictorially a complex number on rectangular (Cartesian) axes, the horizontal axis being used for real numbers and the vertical axis for imaginary numbers.

Revision

1   Draw the following complex numbers on an Argand diagram:

  (a) 2 − j1, (b) −1 + j2

9.2.4 The rectangular and polar forms of complex numbers

We can specify a complex number by either stating its location on an Argand diagram in terms of its *Cartesian* or *rectangular co-ordinates a* and *b*, i.e. as:

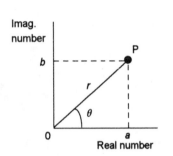

Figure 9.6   *A complex number on an Argand diagram*

$$z = a + jb \qquad [4]$$

or by specifying the length *r* of the line joining the point to the origin and the angle θ the line makes with the positive number axis (Figure 9.6). These are termed its *polar co-ordinates*. The specification in polar co-ordinates can be written as:

$$z = r \angle \theta \qquad [5]$$

In complex number arithmetic, the length *r* of the line OP is called the *modulus* of the complex number and its inclination θ to the real number axis is termed the *argument* of the complex number with the modulus of a complex number *z* being denoted by |*z*|.

9.2.5 Conversion from rectangular form to polar form

Using Pythagoras' theorem we can determine the length OP in Figure 9.6 in terms of the rectangular co-ordinates of the complex number:

$$OP = r = \sqrt{a^2 + b^2} \qquad [6]$$

and, since $\tan \theta = b/a$:

$$\theta = \tan^{-1}\left(\frac{b}{a}\right) \qquad\qquad [7]$$

Example

Determine the polar form of the complex number $2 + j2$.

Using equation [6]:

$$r = \sqrt{a^2 + b^2} = \sqrt{2^2 + 2^2} = 2.8$$

Using equation [7]:

$$\theta = \tan^{-1}\left(\frac{b}{a}\right) = \tan^{-1}\left(\frac{2}{2}\right) = 45°$$

In polar form the complex number could be written as $2.8 \angle 45°$.

Example

Determine the polar form of the complex number $-2 + j2$.

Using equation [6]:

$$r = \sqrt{a^2 + b^2} = \sqrt{(-2)^2 + 2^2} = 2.8$$

Using equation [7]:

$$\theta = \tan^{-1}\left(\frac{b}{a}\right) = \tan^{-1}\left(\frac{2}{-2}\right)$$

If we sketch an Argand diagram (Figure 9.7) for this complex number we can see that the number is in the second quadrant. The argument is thus $-45° + 180° = 135°$. In polar form the complex number could be written as $2.8 \angle 135°$.

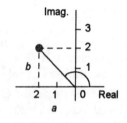

Figure 9.7   *Example*

Revision

2   Express the following complex numbers in polar form:

(a) $3 - j4$, (b) 3, (c) $3 + j4$, (d) $-3 - j4$, (e) $-j4$

9.2.6 Conversion from polar form to rectangular form

Since in Figure 9.6 we have $a = r \cos \theta$ and $b = r \sin \theta$, we can write a complex number z as:

$$z = a + jb = r \cos \theta + jr \sin \theta = r(\cos \theta + j \sin \theta) \qquad [8]$$

Example

Write the complex number $10\angle 60°$ in rectangular form.

Using equation [8]:

$$z = r(\cos \theta + j \sin \theta) = 10(\cos 60° + j \sin 60°) = 5 + j8.7$$

Example

Write the complex number $2\angle 135°$ in rectangular form.

Using equation [8]:

$$z = r(\cos \theta + j \sin \theta) = 2(\cos 135° + j \sin 135°) = -1.4 + j1.4$$

Revision

3   Write the following complex numbers in rectangular form:

(a) $5\angle 60°$, (b) $2\angle -30°$, (c) $4\angle 210°$, (d) $6\angle 310°$.

9.2.7 Addition and subtraction of complex numbers

To add complex numbers in rectangular form we add the real parts together and separately add the imaginary parts together:

$$(a + jb) + (c + jd) = (a + c) + j(b + d) \qquad [9]$$

On an Argand diagram, this method of adding two complex numbers is the same as the vector addition of two vectors using the parallelogram of vectors, the line representing each complex number being treated as a vector (Figure 9.8).

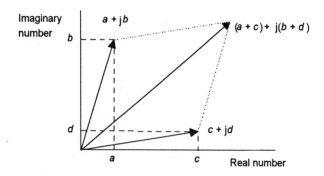

Figure 9.8 *Addition of complex numbers*

To subtract complex numbers we subtract the real parts and separately subtract the imaginary parts:

$$(a + jb) - (c + jd) = (a - c) + j(b - d) \qquad [10]$$

On an Argand diagram, this method of subtracting two complex numbers is the same as the vector subtraction of two vectors. To subtract a vector quantity you reverse its direction and then add it using the parallelogram of vectors (Figure 9.9).

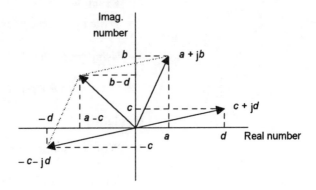

Figure 9.9 *Subtraction of complex numbers*

Example

If $z_1 = 4 + j2$ and $z_2 = 3 + j5$, determine (a) $z_1 + z_2$, (b) $z_1 - z_2$.

(a) With $z_1 = 4 + j2$ and $z_2 = 3 + j5$, then:

$$z_1 + z_2 = (4 + 3) + j(2 + 5) = 7 + j7$$

(b) With $z_1 = 4 + j2$ and $z_2 = 3 + j5$, then:

$$z_1 - z_2 = (4 - 3) + j(2 - 5) = 1 - j3$$

Revision

4   If $z_1 = 2 + j5$, $z_2 = 1 + j3$ and $z_3 = 4 - j2$, determine:

(a) $z_1 + z_2$, (b) $z_2 + z_3$, (c) $z_1 + z_2 + z_3$, (d) $z_1 - z_2$, (e) $z_2 - z_3$

9.2.8 Multiplication of complex numbers

Consider the multiplication of the two complex numbers in Cartesian form, $z_1 = a + jb$ and $z_2 = c + jd$. The product is given by:

$$z_1 z_2 = (a + jb)(c + jd) = ac + j(ad + bc) + j^2 bd$$

and thus, since $j^2 = -1$:

$$z_1 z_2 = ac + j(ad + bc) - bd \qquad\qquad [11]$$

Now consider the multiplication of the two complex numbers in polar form, $z_1 = |z_1|\angle\theta_1$ and $z_2 = |z_2|\angle\theta_2$. Using equation [8] we can write:

$$z_1 = |z_1|(\cos\theta_1 + j\sin\theta_1) \text{ and } z_2 = |z_2|(\cos\theta_2 + j\sin\theta_2)$$

Thus the product is given by:

$$z_1 z_2 = |z_1|(\cos\theta_1 + j\sin\theta_1) \times |z_2|(\cos\theta_2 + j\sin\theta_2)$$

$$= |z_1 z_2| [\cos\theta_1\cos\theta_2 + j(\sin\theta_1\cos\theta_2 + \cos\theta_1\sin\theta_2) + j^2\sin\theta_1\sin\theta_2]$$

$$= |z_1 z_2| [(\cos\theta_1\cos\theta_2 - \sin\theta_1\sin\theta_2) + j(\sin\theta_1\cos\theta_2 + \cos\theta_1\sin\theta_2)]$$

We can use $\cos(A + B) = \cos A \cos B - \sin A \sin B$ and $\sin(A + B) = \sin A \cos B + \cos A \sin B$, to simplify the above equation to:

$$z_1 z_2 = |z_1 z_2|[\cos(\theta_1 + \theta_2) + j\sin(\theta_1 + \theta_2)] \qquad\qquad [12]$$

Hence we can write for the complex numbers in polar form

$$z_1 z_2 = |z_1 z_2|\angle(\theta_1 + \theta_2) \qquad\qquad [13]$$

The magnitude of the product is the product of the magnitudes of the two numbers and its angle is the sum of the angles of the two numbers.

Example

Multiply the two complex numbers $2 - j3$ and $4 + j1$.

Multiplying out, term-by-term:

$$(2 - j3)(4 + j1) = 8 + j2 - j12 - j^2 3 = 8 + j2 - j12 + 3 = 11 - j10$$

Example

Multiply the two complex numbers $3\angle 40°$ and $2\angle 70°$.

Using equation [13]:

$$3\angle 40° \times 2\angle 70° = (3 \times 2)\angle(40° + 70°) = 6\angle 110°$$

Revision

5   Determine the products of the following complex numbers:

   (a) $3 - j5$ and $4 + j2$, (b) $1 - j2$ and $3 + j4$, (c) $2 - j1$ and $3 + j4$

6   Determine the products of the following complex numbers:

   (a) $2\angle 10°$ and $5\angle 30°$, (b) $1\angle 50°$ and $2\angle(-10°)$,

   (c) $3\angle 110°$ and $1\angle 10°$

9.2.9 Complex conjugate

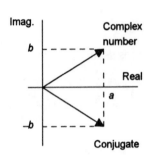

Figure 9.10 *A complex number and its conjugate*

If $z = a + jb$ then the term *complex conjugate* is used for the complex number given by $z^* = a - jb$. The imaginary part of the complex number changes sign to give the conjugate, conjugates being denoted as z^*. Figure 9.10 shows an Argand diagram with a complex number and its conjugate. The complex conjugate is the mirror image of the original complex number.

   Consider now the product of a complex number and its conjugate:

$$zz^* = (a + jb)(a - jb) = a^2 - j^2b = a^2 + b^2 \qquad [14]$$

The product of a complex number and its conjugate is a real number.

Example
What is the conjugate of the complex number $2 + j4$?

The complex conjugate is $2 - j4$.

Example
Determine the product of $2 + j3$ and its conjugate.

The complex conjugate is $2 - j3$ and thus, using equation [14]:

   $(2 + j3)(2 - j3) = 4 + 9 = 13$

Revision

7   Determine the complex conjugates of the following complex numbers:

   (a) $3 - j5$, (b) $-2 + j4$, (c) $-4 - j6$

8   Determine the products of the following complex numbers and their conjugates:

(a) $1 + j2$, (b) $4 - j2$, (c) $-2 + j3$

9.2.10 Division

Consider the division of two complex numbers when in rectangular form; $z_1 = a + jb$ divided by $z_2 = c + jd$:

$$\frac{z_1}{z_2} = \frac{a + jb}{c + jd}$$

To divide one complex number by another we have to convert the denominator into a real number. This can be done by multiplying it by its conjugate. Thus:

$$\frac{z_1}{z_2} = \frac{a + jb}{c + jd} \times \frac{c - jd}{c - jd} = \frac{(a + jb)(c - jd)}{c^2 + d^2} \qquad [15]$$

Now consider the division of the two complex numbers when in polar form; $z_1 = |z_1| \angle \theta_1$ and $z_2 = |z_2| \angle \theta_2$:

$$\frac{z_1}{z_2} = \frac{|z_1|(\cos\theta_1 + j\,\sin\theta_1)}{|z_2|(\cos\theta_2 + j\,\sin\theta_2)}$$

Making the denominator into a real number by multiplying it by its conjugate:

$$\frac{z_1}{z_2} = \frac{|z_1|(\cos\theta_1 + j\,\sin\theta_1)}{|z_2|(\cos\theta_2 + j\,\sin\theta_2)} \times \frac{|z_2|(\cos\theta_2 - j\,\sin\theta_2)}{|z_2|(\cos\theta_2 - j\,\sin\theta_2)}$$

$$= \frac{|z_1|}{|z_2|}\left[\frac{(\cos\theta_1 + j\,\sin\theta_1)(\cos\theta_2 - j\,\sin\theta_2)}{\cos^2\theta_2 + \sin^2\theta_2}\right]$$

But $\cos^2\theta_2 + \sin^2\theta_2 = 1$ and so:

$$\frac{z_1}{z_2} = \frac{|z_1|}{|z_2|}[(\cos\theta_1\cos\theta_2 + \sin\theta_1\sin\theta_2)$$

$$+ j(\sin\theta_1\cos\theta_2 - \cos\theta_1\sin\theta_2)]$$

This can be simplified, using $\cos(A - B) = \cos A \cos B + \sin A \sin B$ and $\sin(A - B) = \sin A \cos A - \cos A \sin B$, to give:

$$\frac{z_1}{z_2} = \frac{|z_1|}{|z_2|}[\cos(\theta_1 - \theta_2) + j\,\sin(\theta_1 - \theta_2)]$$

We can express this as:

$$\frac{z_1}{z_2} = \frac{|z_1|}{|z_2|}\angle(\theta_1 - \theta_2) \qquad [16]$$

Thus to divide two complex numbers in polar form, we divide their magnitudes and subtract their arguments.

Example

Divide $1 + j2$ by $1 + j1$.

$$\frac{1+j2}{1+j1} = \frac{1+j2}{1+j1} \times \frac{1-j1}{1-j1} = \frac{1+j1-j^22}{1-j^2} = \frac{3+j1}{2} = 1.5 + j0.5$$

Example

Divide $4\angle 40°$ by $2\angle 30°$.

$$\frac{4\angle 40°}{2\angle 30°} = \frac{4}{2}\angle(40° - 30°) = 2\angle 10°$$

Revision

9 Determine, in rectangular form, the values of the following:

(a) $\dfrac{2+j3}{3-j4}$, (b) $\dfrac{3+j5}{1+j1}$, (c) $\dfrac{1}{1-j1}$, (d) $\dfrac{2+j3}{5-j2}$,

10 Determine, in polar form, the values of the following:

(a) $\dfrac{4\angle 80°}{2\angle 30°}$, (b) $\dfrac{5\angle 20°}{2\angle 70°}$, (c) $\dfrac{1\angle 80°}{10\angle 70°}$, (d) $\dfrac{2\angle 120°}{4\angle(-20°)}$

9.3 Representing phasors by complex numbers

As indicated in Section 9.2.3, we can describe a line on an Argand diagram of length $|z|$ at an angle θ to the real number axis as the polar form $|z|\angle\theta$ of a complex number (Figure 9.11). A phasor, however, can be regarded as the mathematical representation of an a.c. quantity in polar form. Thus we can treat a phasor as the polar form of a complex number and, as such, we can also represent it by the equivalent rectangular form of complex number, i.e. $a + jb$. Thus:

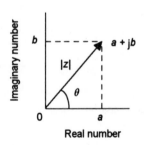

Figure 9.11 *Complex number representation of a phasor*

> A phasor when drawn on an Argand diagram with its base at the origin is represented by a complex number, either in terms of the rectangular components of the complex number $a + jb$ used to describe the point at the tip of the phasor or by the polar representation of the complex number $|z|\angle\theta$ in terms of the length of the line $|z|$ joining it to the origin and its angle θ.

Example

Describe the signal $v = 12\sin(314t + 45°)$ V by a phasor described by a complex number using (a) polar notation and (b) rectangular notion.

(a) The phasor has a magnitude, when expressed as the maximum value, of 12 and phase angle 44°. Thus we can describe it in polar notation as $12\angle 45°$ V. If we used root-mean-square values then we would have $8.49\angle 45°$ V r.m.s.

(b) By using equation [8] we can convert the polar form into the rectangular form as $12 \cos 45° + j12 \sin 45° = 8.49 + j8.49$ V or if we used root-mean-square values as $6 + j6$ V r.m.s.

Revision

11 Describe the following signals by phasors, expressed as the maximum values, in polar and Cartesian forms:

(a) $10 \sin 314t$ V, (b) $2 \sin(314t + 60°)$ A, (c) $5 \sin(6283t + 90°)$ V

9.3.1 Adding or subtracting phasors

Phasors when expressed as complex numbers can be added or subtracted by the methods used for complex number. Such operations are easiest when the complex numbers are in rectangular form.

Example

A circuit has three components in series. If the voltages across each component are described by the phasors 4 V, j2 V and 3 + j4 V, what is the voltage phasor describing the voltage across the three components?

Since the components are in series, the resultant phasor voltage is described by the phasor:

$$\mathbf{V} = 4 + j2 + 3 + j4 = 7 + j6$$

Example

A circuit has two components in series. If the voltages across each component are described by phasors $4\angle 60°$ V and $2\angle 30°$ V, what is the voltage phasor for the voltage across the two components?

For adding complex numbers it is simplest to convert the phasors into rectangular notation. Thus:

$$\mathbf{V} = (4 \cos 60° + j4 \sin 60°) + (2 \cos 30° + j2 \sin 30°)$$

$$= 2 + j3.46 + 1.73 + j1 = 3.73 + j4.46 \text{ V}$$

If we want this phasor in polar notation then, using equations [6] and [7]:

$$V = \sqrt{3.73^2 + 4.46^2} = 5.81$$

$$\phi = \tan^{-1}\frac{4.46}{3.73} = 50°$$

Thus the phasor is $5.81\angle 50°$ V.

Revision

12 Determine the phasor representing the sum of the voltages described by the following phasors, expressing the results in both rectangular and polar notation:

(a) $2 + j3$ V and $1 - j5$ V, (b) 2 V and $-j5$ V,

(c) $4\angle 0°$ V and $3\angle 60°$ V, (d) $5\angle 0°$ V and $10\angle 45°$ V

13 Determine, by means of phasor addition, the sum of the currents $i_1 = 1.5\sin(314t + 30°)$ A and $i_2 = 0.4\sin(314t - 45°)$ A.

9.3.2 Multiplying or dividing phasors

Multiplication or division of complex numbers can be carried out when they are in either rectangular form or polar form, being easiest when they are in polar form. Thus, if we have a voltage across a component described by $\mathbf{V} = V\angle\phi$ and the current by $\mathbf{I} = I\angle\theta$ then the product of the two phasors is:

$$\mathbf{VI} = V\angle\phi \times I\angle\theta = VI\angle(\phi + \theta) \qquad [17]$$

If the voltage and current were in rectangular form, i.e. in the form $\mathbf{V} = a + jb$ and $\mathbf{I} = c + jd$ then the product is:

$$\mathbf{VI} = (a + jb)(c + jd) = (ac - bd) + j(bc + ad) \qquad [18]$$

For division, if we have a voltage across a component described by $\mathbf{V} = V\angle\phi$ and the current by $\mathbf{I} = I\angle\theta$ then:

$$\frac{\mathbf{V}}{\mathbf{I}} = \frac{V\angle\phi}{I\angle\theta} = \frac{V}{I}\angle(\phi - \theta) \qquad [19]$$

If the voltage and current were in rectangular form, i.e. in the form $\mathbf{V} = a + jb$ and $\mathbf{I} = c + jd$ then:

$$\frac{\mathbf{V}}{\mathbf{I}} = \frac{a + jb}{c + jd} = \frac{a + jb}{c + jd} \times \frac{c - jd}{c - jd} = \frac{(a + jb)(c - jd)}{c^2 - d^2} \qquad [20]$$

Example
If phasor $\mathbf{V}$ is $10\angle 30°$ V and phasor $\mathbf{I}$ is $2\angle 45°$ A, determine $\mathbf{VI}$ and $\mathbf{V/I}$.

For the product the situation is similar to that in equation [17], thus:

$$\mathbf{VI} = (10 \times 2)\angle(30° + 45°) = 20\angle75°\ \text{W}$$

For the division the situation is similar to that in equation [19], thus:

$$\frac{\mathbf{V}}{\mathbf{I}} = \frac{10\angle30°}{2\angle45°} = 5\angle(-15°)\ \Omega$$

Example

If phasor **V** is $1 - j2$ V and phasor **I** is $3 + j4$ A, determine **V/I**.

The situation is similar to that in equation [20], thus:

$$\frac{\mathbf{V}}{\mathbf{I}} = \frac{1 - j2}{3 + j4} = \frac{1 - j2}{3 + j4} \times \frac{3 - j4}{3 - j4} = \frac{3 - 8 - j10}{9 + 16} = -0.2 - j0.4$$

Revision

14 Determine the product **VI** if:

(a) phasor **V** is $3 - j4$ V and phasor **I** is $4 + j6$ A,

(b) phasor **V** is $2\angle30°$ V and phasor **I** is $4\angle60°$ A,

(c) phasor **V** is $4\angle(-30°)$ V and phasor **I** is $2\angle60°$ A

15 Determine **V/I** if:

(a) phasor **V** is $2 + j3$ V and phasor **I** is $1 - j2$ A,

(b) phasor **V** is $12\angle80°$ V and phasor **I** is $4\angle60°$ A,

(c) phasor **V** is $10\angle30°$ V and phasor **I** is $2\angle60°$ A

9.4 Impedance *Impedance Z* is defined as the ratio of the phasor voltage across a component to the phasor current through it:

$$Z = \frac{\mathbf{V}}{\mathbf{I}} \qquad\qquad [21]$$

Thus if we have $\mathbf{V} = V\angle\theta$ and $\mathbf{I} = I\angle\phi$ then:

$$Z = \frac{V\angle\theta}{I\angle\phi} = \frac{V}{I}\angle(\theta - \phi) \qquad\qquad [22]$$

Impedance is a complex number but not a phasor since it does not describe a sinusoidally varying quantity. Hence I have not used bold print for it. In some textbooks, however, the convention is used of writing all complex quantities in bold print.

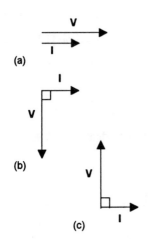

Figure 9.12 *Phasors:*
(a) resistor, (b) capacitor,
(c) inductor

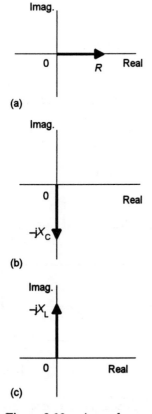

Figure 9.13 *Argand*
diagrams: (a) resistor,
(b) capacitor, (c) inductor

9.4.1 Impedance of circuit elements

For a *pure resistor* the current through it is in phase with the voltage across it (Figure 9.12(a)). Thus for a voltage phasor of $V\angle 0°$ we must have a current phasor of $I\angle 0°$ and so the impedance Z of the circuit element is:

$$Z = \frac{\mathbf{V}}{\mathbf{I}} = \frac{V\angle 0°}{I\angle 0°} = \frac{V}{I}\angle 0° = R\angle 0° \tag{23}$$

The impedance is just the real number V/I which is the resistance R. In rectangular form the impedance is:

$$Z = R \tag{24}$$

There is only a real component and no imaginary component to the impedance. Figure 9.13(a) shows the Argand diagram.

For a *pure capacitance* the current leads the voltage by 90° (Figure 9.12(b)). Thus for a voltage phasor of $V\angle 0°$ we must have a current phasor of $I\angle 90°$ and so the impedance of the circuit element is:

$$Z = \frac{\mathbf{V}}{\mathbf{I}} = \frac{V\angle 0°}{I\angle 90°} = \frac{V}{I}\angle(-90°) = X_C\angle(-90°) \tag{25}$$

The term *capacitive reactance* X_C is used for the ratio of the maximum values of the voltage and current or the root-mean-square values of the voltage and current. In rectangular form the impedance is:

$$Z = -jX_C \tag{26}$$

The impedance is just an imaginary quantity. Figure 9.13(b) shows the Argand diagram. Since $X_C = 1/\omega C$, equation [26] can be written as:

$$Z = -jX_C = -\frac{j}{\omega C} = -\frac{j(j)}{\omega C(j)} = \frac{1}{j\omega C} \tag{27}$$

For a *pure inductance* the current lags the voltage by 90° (Figure 9.13(c)). Thus for a voltage phasor of $V\angle 0°$ we must have a current phasor of $I\angle(-90°)$ and so the impedance of the circuit element is:

$$Z = \frac{\mathbf{V}}{\mathbf{I}} = \frac{V\angle 0°}{I\angle(-90°)} = \frac{V}{I}\angle 90° = X_L\angle 90° \tag{28}$$

The term *inductive reactance* X_L is used for the ratio of the maximum values of the voltage and current or the root-mean-square values of the voltage and current. In rectangular form the impedance is:

$$Z = jX_L \tag{29}$$

The impedance is just an imaginary quantity. Figure 9.5(c) shows the Argand diagram. Since $X_L = \omega L$, equation [29] can be written as:

$$Z = jX_L = j\omega L \tag{30}$$

Example

A capacitor has a capacitance of 0.25 μF. Determine (a) the impedance in polar and rectangular forms and (b) the voltage across it when the current through it is $i = 20 \sin(2000t + 30°)$ mA.

(a) The capacitive reactance is $X_C = 1/\omega C = 1/(2000 \times 0.25 \times 10^{-6}) = 2000\ \Omega$. Thus, in polar form, the impedance is $2000\angle(-90°)\ \Omega$. In rectangular form it is $-j2000\ \Omega$.
(b) Since $\mathbf{V} = \mathbf{I}Z$ then we have $\mathbf{V} = 0.020\angle 30° \times 2000\angle(-90°) = 40\angle(-60°)$ V and so $\mathbf{V} = 40 \sin(2000t - 60°)$ V.

Revision

16 An inductor has an inductance of 8 mH. Determine (a) the impedance in polar and rectangular forms and (b) the voltage across it when the current through it is $10 \sin(2000t + 40°)$ mA.

9.4.2 Components in series

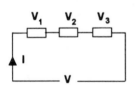

Figure 9.14 *Impedances in series*

If we have impedances connected in series (Figure 9.14), then:

$$\mathbf{V} = \mathbf{V}_1 + \mathbf{V}_2 + \mathbf{V}_3$$

Dividing by the phasor current, the current being the same through each:

$$\frac{\mathbf{V}}{\mathbf{I}} = \frac{\mathbf{V}_1}{\mathbf{I}} + \frac{\mathbf{V}_2}{\mathbf{I}} + \frac{\mathbf{V}_3}{\mathbf{I}}$$

Hence the total impedance Z is the sum of the the three impedances:

$$Z = Z_1 + Z_2 + Z_3 \tag{31}$$

Adding impedances is *not* like adding resistances since impedances are complex quantities and we must take account of both their real and imaginary parts.

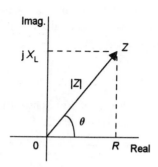

Figure 9.15   *Inductor in
series with a resistor*

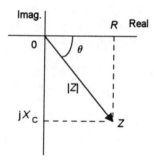

Figure 9.16   *Capacitor in
series with a resistor*

Thus, for a pure inductance in series with a pure resistance we have, using equations [24] and [30] and adding the two impedances:

$$Z = R + jX_L \qquad [32]$$

Figure 9.15 shows the Argand diagram. In polar terms the impedance is $Z = |Z| \angle \theta$.

For a pure capacitance in series with a pure resistance, using equations [24] and [27], by adding the impedances of the two components:

$$Z = R - jX_C \qquad [33]$$

Figure 9.16 shows the Argand diagram. In polar terms the impedance is $Z = |Z| \angle \theta$.

For a pure inductance, pure capacitance and pure resistance in series, using equations [24], [27] and [30], by adding the impedances of the three components:

$$Z = R + jX_L - jX_C = R + j(X_L - X_C) \qquad [34]$$

Figure 9.17 shows the Argand diagram. In polar terms the impedance is $Z = |Z| \angle \theta$.

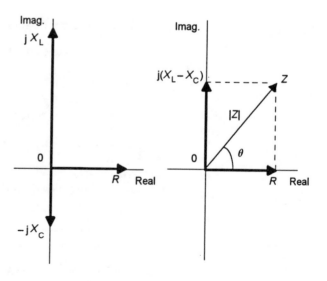

Figure 9.17   *Inductor, capacitor and resistor in series*

Example

Determine the impedance of a resistor with a 100 Ω resistance in series with a capacitor having a capacitive reactance of 5 Ω.

Adding the impedances of the components gives $Z = 100 - j5\ \Omega$.

Example

A resistor with resistance 300 Ω is in series with an inductor of inductance 0.2 H. What will be the current in the circuit when the applied voltage is 20 sin 2000t V?

The reactance of the inductor is $X_L = \omega L = 2000 \times 0.2 = 400\ \Omega$. The impedance of the circuit is thus $R + jX_L = 300 + j400\ \Omega$. In polar notation this is $\sqrt{(300^2 + 400^2)}\angle\tan^{-1}(400/300) = 500\angle 53°\ \Omega$. The current phasor for the circuit $\mathbf{I} = \mathbf{V}/Z = 20\angle 0°/500\angle 53° = 0.04\angle(-53°)$ A and so the current is 0.04 sin(2000t − 53°) A.

Revision

17 Determine the impedance of the following circuits:

(a) a resistor of resistance of 20 Ω in series with an inductor of inductive reactance 100 Ω,

(b) a resistor of resistance of 100 Ω in series with a capacitor of capacitive reactance of 40 Ω,

(c) a resistor of resistance of 10 Ω in series with an inductor of inductive reactance of 20 Ω and a capacitor of capacitive reactance of 5 Ω,

18 An alternating e.m.f. $30\angle 0°$ V is applied to a circuit consisting of a resistance of 200 Ω in series with an inductive reactance of 100 Ω. Determine, in polar notation, the phasor describing the circuit current.

19 An alternating voltage of 12 sin(240t + 45°) V is applied to a circuit consisting of a capacitor of capacitance 2.2 μF in series with a resistor of resistance 3.3 kΩ. Determine in sinusoidal form the current in the circuit and the voltage drops across the resistor and capacitor.

9.4.3 Components in parallel

Consider the parallel connection of impedances (Figure 9.18). For the currents we have:

$$\mathbf{I} = \mathbf{I}_1 + \mathbf{I}_2 + \mathbf{I}_3$$

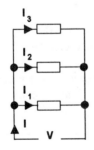

Figure 9.18   *Impedances in parallel*

Dividing by the phasor voltage, the voltage being the same for each impedance:

$$\frac{\mathbf{I}}{\mathbf{V}} = \frac{\mathbf{I_1}}{\mathbf{V}} + \frac{\mathbf{I_2}}{\mathbf{V}} + \frac{\mathbf{I_3}}{\mathbf{V}}$$

Thus the total impedance Z is given by:

$$\frac{1}{Z} = \frac{1}{Z_1} + \frac{1}{Z_2} + \frac{1}{Z_3} \qquad [35]$$

Admittance Y is the reciprocal of impedance and so equation [35] can be written as:

$$Y = Y_1 + Y_2 + Y_3 \qquad [36]$$

With parallel circuits it is often simpler to work in terms of admittances, so avoiding the reciprocals that would occur with impedances. A pure resistor has an admittance of:

$$Y = \frac{1}{R} = G \qquad [37]$$

where G is the conductance. A pure capacitor has an admittance of:

$$Y = \frac{1}{-jX_C} = \frac{j}{-(jX_C)(j)} = \frac{j}{X_C} = jB_C \qquad [38]$$

where B_C is the capacitive susceptance. A pure inductor has an admittance of:

$$Y = \frac{1}{jX_L} = \frac{j}{(jX_L)(j)} = -\frac{j}{X_L} = -jB_L \qquad [39]$$

where B_L is the inductive susceptance. Figure 9.19 shows the Argand diagrams representing the conductance and susceptance elements for a pure resistor, a pure capacitor and a pure inductor.

For a pure inductance in parallel with a pure resistance we have an impedance of R in parallel with one of jX_C and thus, using equation [35]:

$$\frac{1}{Z} = \frac{1}{R} + \frac{1}{jX_L} \qquad [40]$$

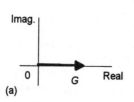

(a)

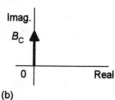

(b)

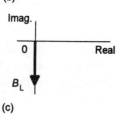

(c)

Figure 9.19　*Argand diagrams: (a) resistor, (b) capacitor, (c) inductor*

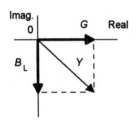

Figure 9.20   *Resistance*
in parallel with inductance

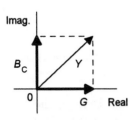

Figure 9.21   *Resistance*
in parallel with capacitance

We can write equation [40] as:

$$Y = \frac{1}{Z} = \frac{1}{R} + \frac{1}{jX_L} = \frac{1}{R} + \frac{j}{(jX_L)(j)} = \frac{1}{R} - j\frac{1}{X_L} = G - jB_C \quad [41]$$

Figure 9.20 shows the Argand diagram for the parallel arrangement of pure resistance and pure inductance.

For a pure capacitance in parallel with a pure resistance we have an impedance of R in parallel with one of $-jX_C$ and so:

$$\frac{1}{Z} = \frac{1}{R} - \frac{1}{jX_C} \quad [42]$$

We can write equation [42] as:

$$Y = \frac{1}{R} - \frac{1}{jX_C} = \frac{1}{R} - \frac{j}{(jX_C)(j)} = \frac{1}{R} + j\frac{1}{X_C} = G + jB_C \quad [43]$$

Figure 9.21 shows the Argand diagram for the parallel arrangement of pure resistance and pure capacitance.

For a parallel arrangement of a pure resistor, a pure inductor and a pure capacitor (Figure 9.22(a)), the Argand diagram for the admittance will be of the form shown in Figure 9.22(b). Thus the admittance of the circuit, when the inductive susceptance is greater than the capacitive susceptance, is:

$$Y = G + j(B_L - B_C) \quad [44]$$

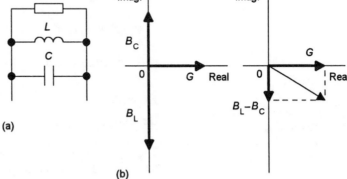

Figure 9.22   *(a) Circuit, (b) Argand diagram*

Example

What is the total impedance of impedances $4\angle 30°$ Ω in parallel with $2\angle(-20°)$ Ω.

Using equation [35]:

$$\frac{1}{Z} = \frac{1}{4\angle 30°} + \frac{1}{2\angle(-20°)} = 0.25\angle(-30°) + 0.5\angle 20$$

$$= 0.25\cos(-30°) + j0.25\sin 30° + 0.5\cos 20° + j0.5\sin 20°$$

$$= 0.686 + j0.296$$

$$= \sqrt{0.686^2 + 0.296^2} \angle \tan^{-1}(0.296/0.686) = 0.747\angle 23.$$

Hence $Z = 1.34\angle(-23.3°)\ \Omega$.

Example

Determine the circuit impedance and the potential difference across the resistor for the circuit shown in Figure 9.23.

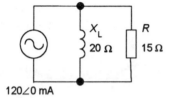

120∠0 mA

Figure 9.23   *Example*

The circuit has an impedance due to a resistance in parallel with an inductance. Thus the impedance is given by equation [35] as:

$$\frac{1}{Z} = \frac{1}{15} + \frac{1}{j20} = \frac{15 + j20}{15 \times j20}$$

$$Z = \frac{j300}{15 + j20} \times \frac{15 - j20}{15 - j20} = \frac{6000 + j4500}{15^2 + 20^2} = 9.6 + j7.2$$

In polar notation this is:

$$Z = \sqrt{9.6^2 + 7.2^2} \angle \tan^{-1}(7.2/9.6) = 12\angle 36.9°$$

The potential difference across this impedance, and hence across the resistance, is thus:

$$\mathbf{V} = Z\mathbf{I} = 12\angle 36.9° \times 120\angle 0°$$

$$= 1440\angle 36.9°\ mV = 1.44\angle 36.9°\ V$$

Example

Determine the total admittance of the circuit shown in Figure 9.24.

The admittance of the resistor is its conductance $G = 1/1000 = 0.0001$ S. The admittance of the inductor is its inductive susceptance $B_L = -j/X_L = -j/(25 \times 10^3 \times 20 \times 10^{-3}) = -j0.002$ S. The admittance of the capacitor is its capacitive susceptance $B_C = j/X_C = j(25 \times 10^3 \times 0.16 \times 10^{-6}) = j0.004$ S. The total admittance is thus:

admittance $= 0.0001 - j0.002 + j0.004 = 0.0001 + j0.002$ S

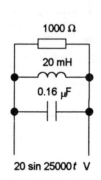

1000 Ω

20 mH

0.16 μF

20 sin 25000t   V

Figure 9.24   *Example*

Revision

20 Determine the impedance of:

(a) a resistor with a resistance of 20 Ω in parallel with an inductor of inductive reactance of 10 Ω,

(b) three components in parallel, a resistor with a resistance of 200 Ω, an inductor with an inductive reactance of 200 Ω and a capacitor with a capacitive reactance of 100 Ω

21 Determine the admittance of:

(a) a resistor of conductance 0.01 S in parallel with a capacitor of susceptance 0.001 S,

(b) a resistor of conductance 0.002 S in parallel with an inductor of susceptance 0.005 S.

22 Determine the total admittance and the total impedance of a circuit consisting of three parallel components, a resistor of resistance 2 Ω, an inductor of inductive reactance 5 Ω and a second inductor of inductive reactance 10 Ω.

23 Determine the total admittance and the total current drawn from the alternating voltage supply of 10 V r.m.s. for a circuit consisting of three parallel components, a resistor of resistance 50 Ω, a capacitor of capacitive reactance 80 Ω and an inductor of inductive reactance 40 Ω.

9.4.4 Series-parallel circuits

The procedure that can be used with a.c. circuits containing series- and parallel-connected components is to reduce such circuits to progressively simpler equivalents by replacing series and parallel combinations of components with their equivalents.

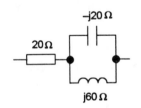

Figure 9.25   *Series-parallel circuit*

Example

Determine the total impedance of the circuit shown in Figure 9.25.

The equivalent impedance of the parallel combination of the capacitor and inductor is given by:

$$\frac{1}{Z_p} = \frac{1}{Z_C} + \frac{1}{Z_L} = \frac{Z_L + Z_C}{Z_C Z_L}$$

$Z_L + Z_C = j60 + (-j20) = j40\ \Omega$; in polar form this is $40\angle 90°$. For $Z_C Z_L$ it is simplest if we use the polar form of the impedances, i.e. $20\angle(-90°)$ and $60\angle 90°$, to give $1200\angle 0°$. Thus:

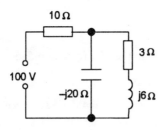

Figure 9.26   *Example*

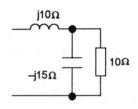

Figure 9.27   *Revision problem 24*

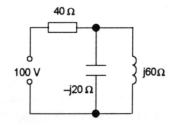

Figure 9.28   *Revision problem 25*

$$z_P = \frac{Z_L Z_C}{Z_L + Z_C} = \frac{1200\angle 0^\circ}{40\angle 90^\circ} = 30\angle(-90^\circ)\ \Omega$$

In polar form this is $-j30\ \Omega$. For the series combination of Z_P and $20\ \Omega$, the total impedance is thus:

$$Z = 20 - j30\ \Omega$$

Example

Determine the current taken from the supply for the circuit shown in Figure 9.26.

For the series resistance and inductance the impedance is $Z_{LR} = 3 + j6\ \Omega = 6.71\angle 63.4^\circ$ W. For the parallel arrangement of this and the capacitor, the impedance Z_P is given by:

$$\frac{1}{Z_p} = \frac{1}{Z_C} + \frac{1}{Z_{LR}} = \frac{Z_{LR} + Z_C}{Z_C Z_{LR}}$$

$$Z_P = \frac{6.71\angle 63.4^\circ \times 20\angle(-90^\circ)}{-j20 + 3 + j6} = \frac{134.2\angle(-26.6^\circ)}{14.3\angle(-77.9^\circ)}$$

$$= 9.38\angle 51.3^\circ = 5.86 + j7.32\ \Omega$$

This parallel arrangement is in series with a resistance and so the total circuit impedance is $10 + 5.86 + j7.32 = 15.86 + j7.32 = 17.46\angle 24.8^\circ\ \Omega$. The circuit current is thus:

$$I = \frac{V}{Z} = \frac{100\angle 0^\circ}{17.46\angle 24.8^\circ} = 5.73\angle(-24.8^\circ)\ A$$

Revision

24 Determine the total impedance of the circuit shown in Figure 9.27.

25 Determine the current taken from the supply for the circuit shown in Figure 9.28.

Problems

1   Represent the following complex numbers on an Argand diagram:

(a) $1 + j1$, (b) $1 - j1$, (c) $-1 - j1$

2   Express the following complex numbers in polar form:

(a) $-4 + j$, (b) $-3 - j4$, (c) 3, (d) $-j6$, (e) $1 + j$, (f) $3 - j2$

3   Express the following complex numbers in rectangular form:

(a) $5\angle 120^\circ$, (b) $10\angle 45^\circ$, (c) $6\angle 180^\circ$, (d) $2.8\angle 76^\circ$, (e) $2\angle 30^\circ$

4   If $z_1 = 3 + j2$ and $z_2 = -2 + j4$, determine the values of:

(a) $z_1 + z_2$, (b) $z_1 - z_2$, (c) z_1z_2, (d) $\frac{1}{z_1}$, (e) $\frac{z_1}{z_2}$

5   Evaluate the following:

(a) $(2 + j3) + (3 - j5)$, (b) $(-4 - j6) + (2 + j5)$, (c) $(2 + j2) - (3 - j5)$,

(d) $(2 + j4) - (1 + j4)$, (e) $4(3 + j2)$, (f) $j2(3 + j5)$,

(g) $(1 - j2)(3 + j4)$, (h) $(2 + j2)(2 - j3)$, (i) $(1 + j2)(4 - j3)$,

(j) $\dfrac{6 + j3}{4 - j2}$, (k) $\dfrac{1}{3 + j2}$, (l) $\dfrac{1 + j1}{1 - j1}$, (m) $\dfrac{3 + j2}{1 - j3}$

6   If $z_1 = 10\angle 20°$, $z_2 = 2\angle 40°$ and $z_3 = 5\angle 60°$, evaluate the following:

(a) z_1z_2, (b) z_1z_3, (c) $\frac{1}{z_1}$, (d) $\frac{1}{z_2}$, (e) $\frac{z_1}{z_2}$, (f) $\frac{z_2}{z_3}$

7   Describe the following signals by phasors written in both polar and rectangular forms, taking the magnitude to represent the maximum value:

(a) $10 \sin(2\pi 50t - \pi/6)$, (b) $10 \sin(314t + 150°)$,

(c) $22 \sin(628t + \pi/4)$

8   Determine, in both rectangular and polar forms, the sum of the following phasors:

(a) $4\angle 0°$ and $3\angle 60°$, (b) $2 + j3$ and $-4 + j4$, (c) $4\angle \pi/3$ and $2\angle \pi/6$

9   If phasors $\mathbf{A}$, $\mathbf{B}$ and $\mathbf{C}$ are represented by $\mathbf{A} = 10\angle 30°$, $\mathbf{B} = 2.5\angle 60°$ and $\mathbf{C} = 2\angle 45°$ determine:

(a) $\mathbf{AB}$, (b) $\mathbf{AC}$, (c) $\mathbf{A}(\mathbf{B} + \mathbf{C})$, (d) $\mathbf{A/B}$, (e) $\mathbf{B/C}$, (f) $\mathbf{C}/(\mathbf{A} + \mathbf{B})$

10   If $v_1 = 10 \sin \omega t$ and $v_2 = 20 \sin(\omega t + 60°)$, what is (a) the phasor describing the sum of the two voltages and (b) its time-domain equation?

11   If the voltage across a component is $5 \sin(314t + \pi/6)$ V and the current through it $0.2 \sin(314t + \pi/3)$ A, what is its impedance?

12   A voltage of 100 V is applied across a circuit of impedance $40 + j30$ Ω, what is, in polar notation, the current taken?

13 A capacitor has a capacitance of 10μF. Determine (a) the impedance of the capacitor in polar and rectangular forms and (b) the voltage across it when the current through it is $i = 0.4 \sin(314t + 40°)$A.

14 An inductor has an inductance of 100 mH. Determine (a) the impedance of the inductor in polar and rectangular forms and (b) the current when the voltage across it is $12 \sin(2000t + 20°)$ V.

15 Determine, in rectangular form, the total impedances of:

(a) 10 Ω in series with $2 - j5$ Ω,

(b) $100\angle30°$ Ω in series with $100\angle60°$ Ω,

(c) $20\angle30°$ Ω in series with $15\angle(-10°)$ Ω,

(d) $20\angle30°$ Ω in parallel with $6\angle(-90°)$ Ω,

(e) 10 Ω in parallel with $-j2$ Ω,

(f) $j40$ Ω in parallel with $j20$ Ω

16 Determine, in rectangular form, the total impedance of:

(a) a resistor with a resistance of 5 Ω in series with an inductor of inductive reactance 2 Ω,

(b) a resistor with a resistance of 50 Ω in series with a capacitor of capacitive reactance 10 Ω,

(c) a resistor with a resistance of 2 Ω in series with an inductor of inductive reactance of 5 Ω and a capacitor of capacitive reactance 4 Ω,

(d) three elements in parallel, a resistor of resistance of 2 Ω, an inductor of inductive reactance of 10 Ω and a capacitor of capacitive reactance 5 Ω,

(e) an inductor of inductive reactance of 500 Ω in parallel with a capacitor of capacitive reactance 100 Ω.

17 Determine, in rectangular form, the total admittance of:

(a) a resistor of resistance 20 Ω in parallel with an inductor of inductive reactance 20 Ω,

(b) a resistor of resistance 10 Ω in parallel with a capacitor of capacitive reactance 500 Ω,

(c) three elements in parallel, a resistor of resistance 200 Ω, an inductor of inductive reactance 800 Ω and a capacitor of capacitance 600 Ω.

18 An alternating voltage of 100 sin 314t V is applied to a circuit consisting of three series elements, a resistor of resistance 800 Ω, a capacitor of capacitive reactance 450 Ω and an inductor of inductive reactance 1250 Ω. Determine the circuit current.

19 An alternating current of 240 sin ωt mA is applied to a circuit consisting of a resistor of resistance 100 Ω in parallel with a capacitor of capacitive reactance 100 Ω. Determine the current through each element.

20 An alternating voltage of 20 sin ωt V is applied to a circuit consisting of a capacitor of capacitive reactance of 10 Ω in series with a parallel arrangement of a resistor of resistance of 10 Ω and an inductor of inductive reactance 10 Ω. Determine the circuit current.

21 An alternating voltage of 20 sin ωt V is applied to a circuit consisting of a coil of resistance 2 Ω and inductive reactance 4 Ω in parallel with a resistor of resistance 5 Ω and a capacitor of capacitive reactance 10 Ω. Determine (a) the current drawn by the circuit, (b) the current through the capacitor.

22 A circuit consists of three components in series, a resistor of resistance 5 Ω, a capacitor of capacitive reactance 6 Ω and an inductor of inductive reactance 10 Ω. A root-mean-square voltage of $10\angle0°$ V is applied to the circuit. Determine the voltage across each element of the circuit.

23 A circuit consists of three components in series, a resistor of resistance 10 Ω, a capacitor of capacitive reactance 0.25 Ω and an inductor of inductive reactance 10 Ω. If the voltage across the inductive reactance is $60\angle40°$ V, what is the voltage applied to the circuit?

24 A coil of resistance 8 Ω and inductive reactance j10 Ω is in parallel with another coil of resistance 7 Ω and inductive reactance j9 Ω. This parallel arrangement of coils is then in series with a resistance of 5 Ω. Determine the total circuit impedance and the current taken from an alternating voltage supply of $100\angle0°$ V.

25 A resistance of 1000 Ω is in parallel with a capacitor of capacitive reactance –j400 Ω. This parallel arrangement is in series with a resistance of 1000 Ω and a capacitor of capacitive reactance –j600 Ω. Determine the total circuit impedance and the current taken from an alternating voltage supply of $100\angle0°$ V.

10 Resonant circuits

10.1 Introduction

The total impedance of any network containing resistance, inductance and capacitance changes when the frequency changes. This is because the reactive component of the impedance changes with frequency and the resistance does not change with frequency. Figure 10.1 shows how the reactance of pure inductors and pure capacitors changes with frequency, the reactance of the pure inductor being given by:

$$X_L = \omega L = 2\pi f L \qquad\qquad [1]$$

and that of the pure capacitor by:

$$X_C = \frac{1}{\omega C} = \frac{1}{2\pi f L} \qquad\qquad [2]$$

The inductive reactance increases with increasing frequency and the capacitive reactance decreases with increasing frequency. Thus it may be possible with some circuit containing both inductance and capacitance to operate at a frequency at which the inductive reactance is just cancelled by the capacitive reactance. The reactive component, i.e. the imaginary part, of the impedance then becomes zero and the total impedance is just that due to resistance. Such a circuit is said to be *resonant*.

This chapter is about series and parallel resonant circuits and the resonant conditions. Such circuits are widely used in that they enable small portions of a spectrum of frequencies to be selected. For example, the tuning of a radio or television receiver is the selection of a small portion of the incoming frequency spectrum.

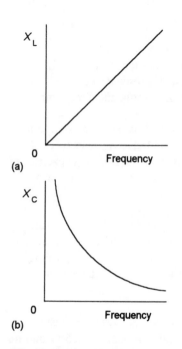

(a)

(b)

Figure 10.1   *Variation with frequency of reactance*

10.2 Resonance with series circuits

Consider a circuit with inductance, resistance and capacitance in series and connected to an alternating voltage supply of frequency f (Figure 10.2(a)). Figure 10.2(b) shows the phasor diagram for the circuit. The circuit impedance Z is, for when X_L is greater than X_C:

$$Z = R + j(X_L - X_C) = R + j\left(\omega L - \frac{1}{\omega C}\right) \qquad [3]$$

In polar notation the impedance is:

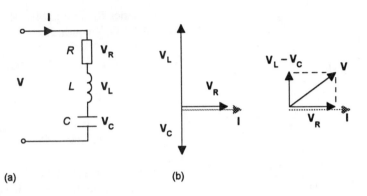

Figure 10.2   *Series circuit*

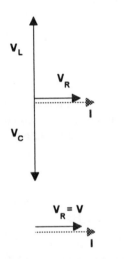

Figure 10.3   *Resonance*

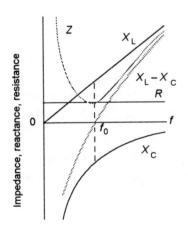

Figure 10.4   *Resonance*

$$Z = \sqrt{R^2 + \left(\omega L - \frac{1}{\omega C}\right)^2} \angle \tan^{-1} \frac{\omega L - 1/\omega C}{R} \qquad [4]$$

The impedance thus depends on the frequency.

There will be a frequency f_0 at which we have $X_L = X_C$. With $X_L = X_C$, since the current through each component is the same, we must have the phasor for the voltage across the inductor V_L with the same magnitude as that of the phasor for the voltage across the capacitor V_C. But the two phasors are in opposite directions and so cancel each other out (Figure 10.3). The circuit voltage is then in phase with the circuit current. When this occurs, $X_L = X_C$ gives:

$$2\pi f_0 L = \frac{1}{2\pi f_0 C}$$

and so:

$$f_0 = \frac{1}{2\pi \sqrt{LC}} \qquad [5]$$

This frequency is known as the *resonant frequency*.

Figure 10.4 shows how the impedance varies with frequency. When $X_L = X_C$, i.e. at the resonant frequency, the impedance is just R. At all other frequencies the impedance is greater than R. The circuit thus has a *minimum impedance* at the resonant frequency.

Since the circuit current $I = V/Z$, the consequence of the circuit impedance being a minimum at the resonant frequency f_0 is that the *circuit current is at its maximum value*, being V/R (Figure 10.5). At frequencies much lower than the resonant frequency and much higher, the impedance is high and so the current is very low.

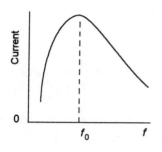

Figure 10.5 *Current variation with frequency*

Example

A series *RLC* circuit has a resonant frequency of 50 Hz, a resistance of 20 Ω, an inductance of 300 mH and a supply voltage of 24 V. Calculate (a) the value of the capacitance, (b) the circuit current at resonance, and (c) the values of the voltages across each component.

(a) The resonant frequency $f_0 = 1/2\pi\sqrt{(LC)}$ and so:

$$C = \frac{1}{4\pi^2 f_0^2 L} = \frac{1}{4\pi^2 \times 50^2 \times 0.300} = 33.8 \ \mu F$$

(b) At resonance $I = V/R = 24/20 = 1.2$ A.
(c) At resonance the voltage across the resistor will be the entire supply voltage of 24 V. The inductive reactance $X_L = 2\pi f L = 2\pi \times 50 \times 0.300 = 94.2$ Ω. At resonance $X_L = X_C$ and so the capacitive reactance is also 94.2 Ω. Hence the value of the voltage across the inductor and across the capacitor is $IX = 1.2 \times 94.2 = 113.0$ V.

Example

A series *RLC* circuit has a resistance of 50 Ω, an inductance of 200 mH and a capacitance of 0.2 μF. Determine the resonant frequency and the current at resonance when the supply voltage is $20\angle 0°$ V.

The resonant frequency is given by equation [4] as:

$$f_0 = \frac{1}{2\pi\sqrt{LC}} = \frac{1}{2\pi\sqrt{0.200 \times 0.2 \times 10^{-6}}} = 795.8 \ Hz$$

At resonance the circuit impedance is the resistance and so 50 Ω. Hence the circuit current is:

$$\mathbf{I} = \frac{\mathbf{V}}{Z} = \frac{20\angle 0°}{50} = 0.4\angle 0° \ A$$

Revision

1   A series *RLC* circuit has a resistance of 40 Ω, an inductance of 10 mH and a capacitance of 0.1 μF. Determine the resonant frequency and the current at resonance when the supply voltage is $10\angle 0°$ V.

2   A coil with a resistance of 20 Ω and an inductance of 75 mH is connected in series with a capacitor of capacitance 40 μF across a supply voltage of $50\angle 0°$ V. Determine the resonant frequency and the circuit current at resonance.

3   A coil with a resistance of 10 Ω and an inductance of 50 mH is connected in series with a capacitor of capacitance 12 μF across a

supply voltage of $20\angle 0°$ V. Determine the resonant frequency and the voltage across the capacitor at resonance.

10.2.1 Q factor

In a series RLC circuit, at resonance the size of the voltage across the inductance is IX_L and that across the capacitor is IX_C. They can be both much larger than the voltage across the resistor. A factor known as the *Q-factor* or 'quality factor' is used to indicate the voltage magnification across either the inductor or capacitor at resonance. It can be defined as:

$$Q\text{-factor} = \frac{\text{voltage across } L \text{ or } C \text{ at resonance}}{\text{voltage across } R \text{ at resonance}} \qquad [6]$$

Thus in terms of the voltage across the inductance:

$$Q\text{-factor} = \frac{IX_L}{IR} = \frac{X_L}{R} = \frac{2\pi f_0 L}{R} \qquad [7]$$

or, in terms of the voltage across the capacitance:

$$Q\text{-factor} = \frac{IX_C}{IR} = \frac{X_C}{R} = \frac{1}{2\pi f_0 CR} \qquad [8]$$

Since $f_0 = 1/2\pi\sqrt{(LC)}$ we can write for either equation [7] or [8]:

$$Q\text{-factor} = \frac{1}{R}\sqrt{\frac{L}{C}} \qquad [9]$$

Equation [6] stems from a more fundamental definition of Q-factor as:

$$Q\text{-factor} = 2\pi \times \frac{\text{max. energy stored in a component}}{\text{energy dissipated per cycle}} \qquad [10]$$

For a series RLC circuit, at resonance the average power dissipated is just that dissipated in the resistance and so is $I^2_{\text{r.m.s}}R = (I_{\text{max}}/\sqrt{2})^2 R = I^2_{\text{max}}R/2$. The maximum energy stored in a capacitor is $\frac{1}{2}CV^2_{\text{max}}$. Since $V_{\text{max}} = I_{\text{max}}X_C$, then the maximum energy stored in a capacitor is $\frac{1}{2}CI^2_{\text{max}}X^2_C$. Thus equation [10] gives:

$$Q\text{-factor} = 2\pi \frac{\frac{1}{2}CI^2_{\text{max}}X^2_C}{\frac{1}{2}I^2_{\text{max}}R} = \frac{X_C}{R} \qquad [11]$$

Since the size of the voltage across the inductance is IX_L and that across the capacitor is IX_C then the above gives the equation [6] definition.

The Q-factor in giving a measure of the voltage magnification at resonance, also gives a measure of the selectivity of the circuit. The

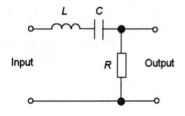

Figure 10.6   *Bandpass filter*

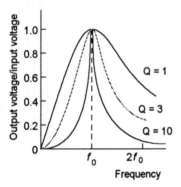

Figure 10.7   *Bandpass filter*

greater the voltage magnification at the resonant frequency the more selective the circuit is of the band of frequencies centred on the resonant frequency from the entire spectrum of frequencies. A high Q-factor thus means a high selectivity.

As an illustration, consider the series RLC circuit when used as a bandpass filter. A bandpass filter is one that is used to pass a band of frequencies out of a spectrum of frequencies. Figure 10.6 shows the circuit, the output being the voltage across the resistor. At the resonant frequency the reactance of the inductor cancels out the reactance of the capacitor and so the entire input voltage appears across the resistor. At higher frequencies the reactance of the inductor is large and that of the capacitor small, thus the circuit impedance increases and so the circuit current decreases and the voltage drop across the resistor is smaller. At lower frequencies the reactance of the capacitor is large and that of the inductor small, thus the circuit impedance increases and so the circuit current decreases and the voltage drop across the resistor is smaller. Figure 10.7 shows how the voltage across the resistor changes with frequency for different Q-factors. A high Q-factor gives a filter with a smaller bandwidth and so greater selectivity.

Example

A series RLC circuit has a resonant frequency of 50 Hz, a resistance of 20 Ω, an inductance of 300 mH and a supply voltage of 24 V. Calculate the Q-factor of the circuit.

This uses the data given for an earlier example in this chapter. The resonant frequency $f_0 = 1/2\pi\sqrt{(LC)}$ and so:

$$C = \frac{1}{4\pi^2 f_0^2 L} = \frac{1}{4\pi^2 \times 50^2 \times 0.300} = 33.8 \ \mu F$$

At resonance $I = V/R = 24/20 = 1.2$ A. The inductive reactance $X_L = 2\pi f L = 2\pi \times 50 \times 0.300 = 94.2 \ \Omega$. At resonance $X_L = X_C$ and so the capacitive reactance is also 94.2 Ω. Hence the value of the voltage across the inductor and across the capacitor is $IX = 1.2 \times 94.2 = 113.0$ V. The Q-factor is voltage across L or C divided by the voltage across R and so is 113.0/24 = 4.7.

Alternatively we could have used equation [9]:

$$Q\text{-factor} = \frac{1}{R}\sqrt{\frac{L}{C}} = \frac{1}{20}\sqrt{\frac{0.300}{33.8 \times 10^{-6}}} = 4.7$$

Revision

4   A series RLC circuit has an inductance of 10 mH and a capacitance of 0.1 μF. What will the resistance have to be if a Q-factor of 15 is to be obtained?

10.3 Resonance with parallel circuits

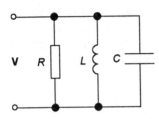

Figure 10.8   *Parallel RLC circuit*

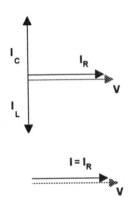

Figure 10.10   *Phasors for the currents at resonance*

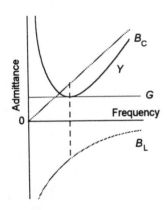

Figure 10.11   *Circuit admittance*

Consider a parallel circuit containing a pure resistor, a pure inductor and a pure capacitor connected as shown in Figure 10.8. The circuit current **I** is the sum of the currents $\mathbf{I_R}$ through the resistor, $\mathbf{I_L}$ through the inductor and $\mathbf{I_C}$ through the capacitor. Figure 10.9 shows the phasor diagram. The size of the current through the inductor is V/X_L and that through the capacitor is V/X_C. They will be the same size and cancel each other out when $X_L = X_C$. The total current is then just $\mathbf{I_R}$ and is in phase with the voltage (Figure 10.10). This is the *resonance* condition.

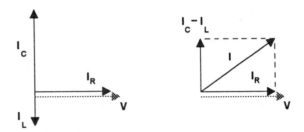

Figure 10.9   *Phasors for the currents*

The admittance of the resistor is just the conductance $G = 1/R$, that of the inductor is just the susceptibility $B_L = 1/\omega L$ and that of the capacitor is just the susceptibility $B_C = \omega C$. The total circuit admittance Y is thus:

$$Y = \frac{1}{R} - \frac{j}{\omega L} + j\omega C = \frac{1}{R} + j\left(\omega C - \frac{1}{\omega L}\right) \qquad [12]$$

In polar notation:

$$Y = \sqrt{\left[\frac{1}{R^2} + \left(\omega C - \frac{1}{\omega L}\right)^2\right]} \angle \tan^{-1} R\left(\omega C - \frac{1}{\omega L}\right) \qquad [13]$$

Figure 10.11 shows how the various terms in the above equation and their sum vary with frequency.

When $X_C = X_L$, i.e. $1/\omega C = \omega L$ and so $\omega C = 1/\omega L$, then the admittance is just $1/R$ and the phase zero. This is the resonance condition and is the condition of *minimum admittance* and hence *maximum impedance*. Since $\mathbf{I} = \mathbf{V}/Z$ then the condition of maximum impedance means that the current is a minimum.

The frequency at which this condition occurs, i.e. the resonant frequency f_0, is thus given by $\omega C = 1/\omega L$ as:

$$2\pi f_0 C = \frac{1}{2\pi f_0 L}$$

and hence:

$$f_0 = \frac{1}{2\pi \sqrt{LC}} \qquad [14]$$

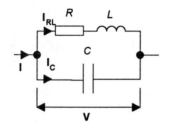

Figure 10.12   *RL in parallel*
with C

The resonant frequency for this form of parallel circuit is thus given by the same equation as that for the series *RLC* circuit.

The parallel circuit described above is an ideal one and involves only pure components. In practice an inductor will invariably have resistance in addition to its inductance. Thus a more practical form of parallel circuit is that shown in Figure 10.12 and has a coil, with both inductance and resistance, in parallel with a capacitor which can generally be reasonably assumed to be pure. The admittance of the inductive branch of the circuit is:

$$Y_{RL} = \frac{1}{R+j\omega L} = \frac{1}{R+j\omega L} \times \frac{R-j\omega L}{R-j\omega L} = \frac{R-j\omega L}{R^2+\omega^2 L^2}$$

and the admittance of the capacitive branch is:

$$Y_C = j\omega C$$

The total admittance of the circuit *Y* is:

$$Y = Y_{RL} + Y_C = \frac{R-j\omega L}{R^2+\omega^2 L^2} + j\omega C$$

$$= \frac{R}{R^2+\omega^2 L^2} + j\left(\omega C - \frac{\omega L}{R^2+\omega^2 L^2}\right) \qquad [15]$$

In polar form this is:

$$Y = \sqrt{\left(\frac{R}{R^2+\omega^2 L^2}\right)^2 + \left(\omega C - \frac{\omega L}{R^2+\omega^2 L^2}\right)^2}$$

$$\angle \tan^{-1}\left[\left(\omega C - \frac{\omega L}{R^2+^2 L^2}\right)\left(\frac{R^2+\omega^2 L^2}{R}\right)\right] \qquad [16]$$

Resonance occurs when the imaginary term in equation [15] is zero, this being the same as stating that it occurs when **I** and **V** are in phase and the phase is zero in equation [16]. Thus resonance occurs at the frequency $\omega_0 = 2\pi f_0$ when:

$$\omega_0 C = \frac{\omega_0 L}{R^2+\omega_0^2 L^2}$$

$$R^2 C + \omega_0^2 L^2 C = L$$

$$\omega_0^2 = \frac{L-R^2 C}{L^2 C} = \frac{1}{LC} - \frac{R^2}{L^2}$$

$$f_0 = \frac{1}{2\pi}\sqrt{\frac{1}{LC} - \frac{R^2}{L^2}} \qquad [17]$$

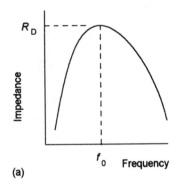

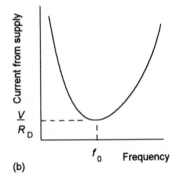

(a)

(b)

Figure 10.13   *Variation of*
(a) impedance, (b) current
with frequency

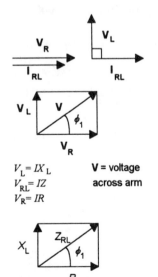

$V_L = IX_L$
$V_{RL} = IZ$
$V_R = IR$

V = voltage
across arm

Figure 10.14   *RL arm of*
circuit

Often R^2/L^2 is negligible when compared with $1/LC$ and the equation reduces to the form:

$$f_0 \approx \frac{1}{2\pi \sqrt{LC}}$$ [18]

At the resonant frequency, the admittance is given by equation [16] as:

$$Y = \frac{R}{R^2 + \omega_0^2 L^2}$$ [19]

and, substituting for ω_0^2, using equation [17], gives:

$$Y = \frac{R}{R^2 + \left(\frac{1}{LC} - \frac{R^2}{L^2}\right)L^2} = \frac{R}{L/C}$$ [20]

and thus the impedance at the resonant frequency is L/RC. This impedance has no imaginary term and is purely real and hence resistive. It is termed the *dynamic resistance* R_D.

$$R_D = \frac{L}{RC}$$ [21]

A consequence of the above equation is that if $R = 0$ then R_D equals infinity. This implies zero current at resonance when $R = 0$.

Figure 10.13 shows how (a) the impedance and (b) the current drawn from the supply by the circuit varies with frequency. The impedance is at a maximum value of R_D and the current at a minimum value of V/R_D at the resonant frequency. For this reason the parallel resonant circuit is known as the *rejecter circuit* since it 'rejects current' at the resonant frequency.

We can derive the above equations by considering the phasor diagrams. For the inductive arm, the effect of having a resistance in series with an inductance is to give the phasors shown in Figure 10.14. The outcome is a current of V/Z_{RL} which is at a phase angle of ϕ_1 to the circuit current. Figure 10.15 shows the current phasors for the parallel circuit. Resonance occurs when the magnitude of the current through the capacitor I_C is equal to $I_{RL} \sin \phi_1$. One way of describing this equality is that the reactive current through the capacitor equals the reactive current through the RL arm, the current through the RL arm having a reactive component which is at 90° to the voltage and a resistive component which is in phase with the voltage. Since $I_C = V/X_C$, $I_{RL} = V/Z_{RL}$ and $\sin \phi_1 = X_L/Z_{RL}$, where X_C is the reactance of the capacitor and Z_{RL} is the impedance of the RL combination, then at resonance:

$$I_C = I_{RL} \sin \phi_1$$

and so:

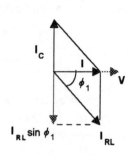

Figure 10.15   *Resonance*

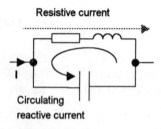

Resistive current

Circulating
reactive current

Figure 10.16   *RL in parallel*

with C at resonance

$$\frac{V}{X_C} = \frac{V}{Z_{RL}} \frac{X_L}{Z_{RL}}$$

and hence:

$$Z_{RL}^2 = X_L X_C$$

Since $X_L = \omega_0 L$, $X_C = 1/\omega_0 C$ and $Z_{RL} = \sqrt{[R^2 + (\omega_0 L)^2]}$, then the above equation can be written as:

$$R^2 + (\omega_0 L)^2 = \omega_0 L \times \frac{1}{\omega C} = \frac{L}{C}$$

and so:

$$f_0 = \frac{1}{2\pi} \sqrt{\frac{1}{LC} - \frac{R^2}{L^2}}$$

which is the equation derived earlier.

At resonance the current I_C can be much larger than the current I drawn from the supply. We can think of this reactive current as circulating round the RL-C circuit with only the resistive component of the current through RL passing through the system (Figure 10.16).

Example

A circuit consists of three components in parallel, a pure resistor of resistance 10 kΩ, an inductor of inductance 0.25 mH and a capacitor of capacitance 50 pF. Determine (a) the resonant frequency, (b) the circuit impedance at resonance and (c) the current taken at resonance by the circuit when a voltage of $10\angle 0°$ V is applied.

(a) The resonant frequency for the parallel circuit is given by equation [14] as:

$$f_0 = \frac{1}{2\pi \sqrt{LC}} = \frac{1}{2\pi \sqrt{0.25 \times 10^{-3} \times 50 \times 10^{-12}}} = 1.42 \text{ MH}$$

(b) At resonance the total circuit impedance is the resistance and so 10 kΩ.
(c) The current taken is:

$$I = \frac{V}{Z} = \frac{10\angle 0°}{10 \times 10^3} = 1\angle 0° \text{ mA}$$

Example

A coil of resistance 10 Ω and inductance 50 mH is connected in parallel with a capacitor of capacitance 10 μF. Determine (a) the resonant frequency, (b) the impedance at resonance and (c) the current taken by the circuit at resonance when a voltage of $5\angle 0°$ V is applied.

(a) Using equation [17]:

$$f_0 = \frac{1}{2\pi}\sqrt{\frac{1}{LC} - \frac{R^2}{L^2}} = \frac{1}{2\pi}\sqrt{\frac{1}{0.050 \times 10 \times 10^{-6}} - \frac{10^2}{0.050^2}}$$

$$= 223 \text{ Hz}$$

Neglecting the R^2/L^2 term gives the answer 225 Hz.

(b) At resonance the impedance is the dynamic resistance and given by equation [21] as:

$$R_D = \frac{L}{RC} = \frac{0.050}{10 \times 10 \times 10^{-6}} = 500 \ \Omega$$

(c) The current at resonance is given by:

$$I = \frac{V}{Z} = \frac{5\angle 0^\circ}{500} = 10\angle 0^\circ \text{ mA}$$

Revision

5   A circuit consists of three components in parallel, a pure resistor of resistance 100 Ω, an inductor of inductance 4 mH and a capacitor of capacitance 1.6 μF. Determine (a) the resonant frequency, (b) the circuit impedance at resonance and (c) the current taken at resonance by the circuit when a voltage of $6\angle 0^\circ$ V is applied.

6   A coil of resistance 10 Ω and inductance 5 mH is connected in parallel with a capacitor of capacitance 0.25 μF. Determine (a) the resonant frequency, (b) the impedance at resonance and (c) the current taken by the circuit at resonance when a voltage of $10\angle 0^\circ$ V is applied.

10.3.1 Q factor

The *Q-factor* of the parallel circuits considered in the previous section at resonance can be defined as being the current magnification produced by a component (or derived from the basic equation [10]) and so:

$$Q\text{-factor} = \frac{I_c}{I} \qquad\qquad [22]$$

where I_c is the size of the current through the component and I that taken from the supply by the circuit. For the resistance, inductance and capacitance each being in separate parallel arms, the current through the capacitor is V/X_C and, since at resonance the total impedance of the circuit is just the resistance R, the current taken from the supply is V/R. Thus:

$$Q\text{-factor} = 2\pi f_0 C \qquad\qquad [23]$$

If, instead of the capacitor we had consider the Q-factor for the inductor, the current through the inductor is V/X_L and, since at resonance the total impedance of the circuit is just the resistance R, the current taken from the supply is V/R. Thus:

$$Q\text{-factor} = \frac{R}{2\pi f_0 L} \qquad [24]$$

At resonance, since $X_L = X_C$, equations [23[and [24] give the same answer.

Consider the parallel circuit involving a coil, having resistance and inductance, in parallel with a capacitor. For the capacitor, the current through it is V/X_C and, since at resonance the total impedance of the circuit is just the dynamic resistance R_D, the current taken from the supply is V/R_D. Thus:

$$Q\text{-factor} = 2\pi f_0 C R_D$$

Since $R_D = L/RC$ (equation [21]), then we have:

$$Q\text{-factor} = \frac{2\pi f_0 L}{R} \qquad [25]$$

Since at resonance, the reactance element of the coil is equal to the reactance of the capacitor, equation [25] can also be written as:

$$Q\text{-factor} = \frac{1}{2\pi f_0 CR} \qquad [26]$$

Example

A circuit has a coil of resistance 100 Ω and inductance 50 mH in parallel with a capacitor of 0.01 μF. Determine the resonant frequency, the dynamic resistance and the Q-factor.

Using equation [17]:

$$f_0 = \frac{1}{2\pi} \sqrt{\frac{1}{LC} - \frac{R^2}{L^2}} = \frac{1}{2\pi} \sqrt{\frac{1}{0.050 \times 0.01 \times 10^{-6}} - \frac{100^2}{0.050^2}}$$

Hence the resonant frequency is 7111 Hz. If, instead of the above equation, we had used the approximate version of the equation the answer would have been 7118 Hz. The dynamic resistance is given by equation [21] as:

$$R_D = \frac{L}{CR} = \frac{0.050}{0.01 \times 10^{-6} \times 100} = 50 \text{ k}\Omega$$

The Q-factor is given by equation [25] as:

$$Q\text{-factor} = \frac{2\pi f_0 L}{R} = \frac{2\pi \times 7111 \times 0.050}{100} = 22.$$

Revision

7 A circuit consists of a coil of resistance 10 Ω and inductance 1 mH in parallel with a capacitor of capacitance 0.001 μF. Determine the resonant frequency and the Q-factor.

Problems

1 A series RLC circuit has a resistance of 10 Ω, an inductance of 100 mH and a capacitance of 0.2 μF. Determine the resonant frequency and the current at resonance when the supply voltage is $12\angle 0°$ V.

2 A series RLC circuit has a resistance of 10 Ω, an inductance of 100 mH and a capacitance of 2 μF. What is the resonant frequency of the circuit?

3 A series RLC circuit has a resistance of 50 Ω, an inductance of 200 mH and a capacitance of 0.2 μF. Determine the voltages across each component at resonance if the supply voltage is $15\angle 0°$.

4 A series RLC circuit resonates at a frequency of 100 Hz. If the resistance is 20 Ω and the inductance 300 mH, what is (a) the circuit capacitance, (b) the Q-factor of the circuit?

5 A series RLC circuit has a resistance of 4 Ω, an inductance of 60 mH and a capacitance of 30 μF. Determine (a) the resonant frequency and (b) the Q-factor of the circuit.

6 A coil with a resistance of 10 Ω and an inductance of 50 mH is in parallel with a capacitance of 0.01 μF. Determine the resonant frequency.

7 A coil with a resistance of 5 Ω and an inductance of 50 mH is in parallel with a capacitance of 0.1 μF and a voltage supply of 100 V r.m.s., variable frequency, is applied. Determine the resonant frequency, the dynamic resistance, the current drawn at resonance and the Q-factor.

8 A coil with a resistance of 10 Ω and an inductance of 120 mH is in parallel with a capacitance of 60 μF and a voltage supply of 100 V r.m.s., variable frequency, is applied. Determine the resonant frequency, the dynamic resistance, the current drawn at resonance and the Q-factor.

11 Complex waveforms

11.1 Introduction

The alternating waveforms considered in Chapters 8, 9 and 10 were all assumed to be sinusoidal. This will not always be the case. For example, many voltages which might initially have been sinusoidal have their waveforms 'distorted' by being applied to some non-linear device and thus we need to be able to considered the behaviour of this waveform with an electrical circuit. In other cases we might have a rectangular waveform rather than a sinusoidal one. A waveform which is not sinusoidal is termed a *complex waveform*. This chapter is a basic consideration of circuit analysis with non-sinusoidal waveforms.

11.1.1 Principle of superposition

The basic principle we will be using with circuit analysis using complex waveforms is the *principle of superposition*. When this is applied to circuits termed *linear circuits*, then because the current is proportional to the voltage, if a voltage v_1 gives a current i_1 and a voltage v_2 gives a current i_2 then a voltage equal to the sum of these voltages, i.e. $(v_1 + v_2)$, will give a current equal to the sum of the currents, i.e. $(i_1 + i_2)$. For example, consider the situation where we first apply a d.c. voltage of 2 V to a circuit and determine the current, then apply a d.c. voltage of 3 V and determine the current. If we then applied a voltage of 5 V to the circuit, the current would be the same as the sum of the currents due to the 2 V and the 3 V when separately applied. As another example, if we first apply a sinusoidal voltage of $2 \sin \omega t$ V to a circuit and determine the current and then apply a sinusoidal voltage of $3 \sin \omega t$ V to the circuit and determine the current, then the current we obtain when applying a voltage of $5 \sin \omega t$ V is the same as the sum of the current obtained when $2 \sin \omega t$ V and $3 \sin \omega t$ V were separately applied.

11.2 The Fourier series

A waveform which varies sinusoidally with time can be represented by an equation of the form:

$$v_1 = A \sin \omega t \qquad\qquad [1]$$

where ω is the angular frequency and equal to $2\pi f$, f being the frequency, v_1 is the value of this waveform at time t and A is the maximum value of the waveform, i.e. its amplitude. Such a waveform repeats itself every 2π radians, i.e. 360°. If the frequency is doubled, with the maximum value unchanged, then equation [1] becomes:

$$v_2 = A \sin 2\omega t \qquad [2]$$

If the frequency is trebled, with the maximum value unchanged, we have:

$$v_3 = A \sin 3\omega t \qquad [3]$$

Suppose that, in addition to doubling and trebling the frequency, we also change the maximum values of the waveforms. Equations [1], [2] and [3] can then be written as:

$$v_1 = A_1 \sin 1\omega t \qquad [4]$$

$$v_2 = A_2 \sin 2\omega t \qquad [5]$$

$$v_3 = A_3 \sin 3\omega t \qquad [6]$$

The above equations all describe waveforms that have started off with v being zero at time $t = 0$. When this is not the case then equations [4], [5] and [6] become:

$$v_1 = A_1 \sin(1\omega t + \phi_1) \qquad [7]$$

$$v_2 = A_2 \sin(2\omega t + \phi_2) \qquad [8]$$

$$v_3 = A_3 \sin(3\omega t + \phi_3) \qquad [9]$$

How does this help with describing a complex waveform? Well, Jean Baptiste Fourier in 1822 proposed that any periodic waveform can be made up of a combination of sinusoidal waveforms, i.e.

$$v = A_1 \sin(1\omega t + \phi_1) + A_2 \sin(2\omega t + \phi_2) + A_3 \sin(3\omega t + \phi_3) + \dots \quad [10]$$

We can also include a d.c. component A_0 to give:

$$v = A_0 + A_1 \sin(1\omega t + \phi_1) + A_2 \sin(2\omega t + \phi_2) + A_3 \sin(3\omega t + \phi_3) + \dots$$
$$[11]$$

This is termed the *Fourier* series. The waveform element with the $1\omega t$ frequency is called the *fundamental frequency* or the *first harmonic*, the element with the $2\omega t$ frequency the *second harmonic*, the element with the $3\omega t$ the third harmonic, and so on.

There is an alternative, simpler, way of writing equation [11]. Since $\sin(A + B) = \sin A \cos B + \cos A \sin B$ we can write:

$$A_1 \sin(1\omega t + \phi_1) = A_1 \sin \phi_1 \cos 1\omega t + A_1 \cos \phi_1 \sin 1\omega t$$

If we represent the non-time varying terms $A_1 \sin \phi_1$ by a constant a_1 and $A_1 \cos \phi_1$ by b_1, then:

$$A_1 \sin(1\omega t + \phi_1) = a_1 \cos 1\omega t + b_1 \sin 1\omega t$$

Likewise we can write:

$$A_2 \sin(2\omega t + \phi_2) = a_2 \cos 2\omega t + b_2 \sin 2\omega t$$

$$A_3 \sin(3\omega t + \phi_3) = a_3 \cos 3\omega t + b_3 \sin 3\omega t$$

and so on. Thus equation [11], for convenience we choose to write $\frac{1}{2}a_0$ for A_0, can be written as:

$$v = \frac{1}{2}a_0 + a_1 \cos 1\omega t + a_2 \cos 2\omega t + a_3 \cos 3\omega t + \ldots$$
$$+ b_1 \sin 1\omega t + b_2 \sin 2\omega t + b_3 \sin 3\omega t + \ldots \qquad [12]$$

Fourier series: any periodic waveform can be represented by a constant d.c. signal term plus terms involving sinusoidal waveforms of multiples of a basic frequency.

Revision

1   What harmonics are present in the waveform given by $v = 1.0 - 0.67 \cos 2\omega t - 0.13 \cos 4\omega t$?

11.2.1 Graphical synthesis of a complex waveform

When we have two waveforms we can obtain the waveform describing their sum by adding the values of the two, ordinate by ordinate. Figure 11.1 illustrates this with the addition of a d.c. signal $v = 1$ and a sinusoidal waveform $v = 1 \sin \omega t$ to give the waveform for $1 + 1 \sin \omega t$.

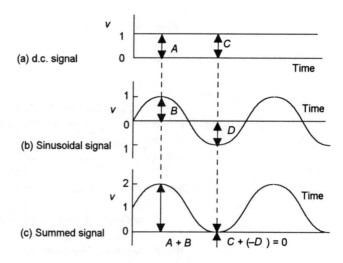

Figure 11.1   *Addition of two waveforms*

Consider the waveform produced by having just sine terms with the fundamental and the third harmonic and $a_3 = a_1/3$, i.e.

$$v = a_1 \sin 1\omega t + \tfrac{1}{3}a_1 \sin 3\omega t \qquad [13]$$

Figure 11.2 shows graphs of the two terms and the waveform obtained by adding the two, ordinate by ordinate.

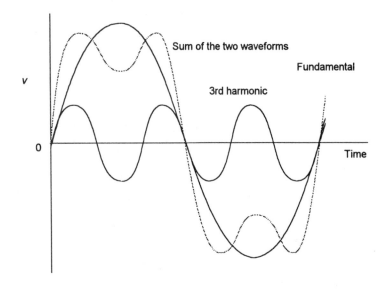

Figure 11.2 *Adding two waveforms*

The result of adding the two waveforms is something that begins to look a bit like a square waveform. A better approximation to a square waveform is given by adding more terms:

$$v = a_1 \sin 1\omega t + \tfrac{1}{3}a_1 \sin 3\omega t + \tfrac{1}{5}a_1 \sin 5\omega t + \tfrac{1}{7}a_1 \sin 7\omega t + ... \qquad [14]$$

If we add to the waveform a d.c. term then we can shift the sum graph up or down accordingly. Thus if we add $0.79a_1$, i.e. we have:

$$v = 0.79a_1 + a_1 \sin 1\omega t + \tfrac{1}{3}a_1 \sin 3\omega t + \tfrac{1}{5}a_1 \sin 5\omega t$$
$$+ \tfrac{1}{7}a_1 \sin 7\omega t + ... \qquad [15]$$

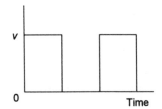

Figure 11.3   *Rectangular waveform*

we obtain a rectangular waveform which approximates to a periodic sequence of pulses (Figure 11.3). Figure 11.4 shows the addition of such a term to the waveform giving Figure 11.2. Note that the average value of this waveform over one cycle is 0.79. The term $\tfrac{1}{2}a_0$ in the Fourier series (equation [12]) thus represents the average value of the waveform over a cycle.

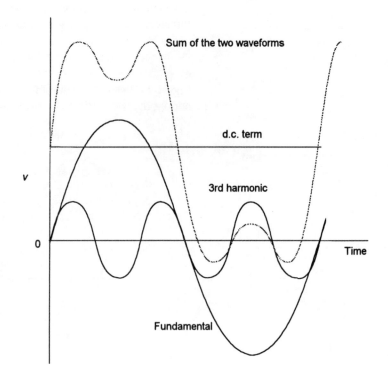

Figure 11.4 *Adding waveforms*

Revision

2   By constructing a graph, determine the waveform for

    (a) $v = 10 \sin \omega t + 3 \cos 3\omega t$,

    (b) $v = 5 + 10 \sin \omega t + 3 \cos 3\omega t$.

11.2.2 Waveforms containing just odd- or even-order harmonics

It is often possible by considering the symmetry of successive half-cycle waves within a waveform to recognise whether it will contain odd or even harmonics.

> Any complex waveform which has a negative half cycle which is just the positive cycle inverted will contain only odd harmonics, such a form of symmetry being termed *half-wave inversion*. Waveforms which repeat themselves after each half-cycle of the fundamental frequency will have just even harmonics, such a form of symmetry being termed *half-wave repetition*.

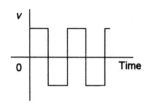

Figure 11.5   *Waveform with identical positive and negative half-cycles*

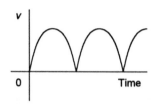

Figure 11.6   *Waveform which repeats every half-cycle*

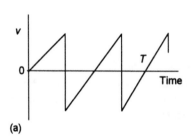

(a)

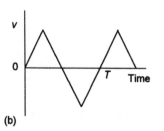

(b)

Figure 11.7   *Revision problem 3*

Thus Figure 11.5 shows a waveform which has negative half cycles which are just the positive half-cycles inverted and so does not contain any even harmonics. Figure 11.6 show a waveform which repeats itself after each half-cycle and so has just even harmonics.

We can see why the above statements occur by considering the conditions that are necessary for a Fourier series to give the required symmetry. Thus if we have the series (equation [12]) describing the waveform at time t:

$$v \text{ at } t = \tfrac{1}{2}a_0 + a_1 \cos 1\omega t + a_2 \cos 2\omega t + a_3 \cos 3\omega t + \ldots$$
$$+ b_1 \sin 1\omega t + b_2 \sin 2\omega t + b_3 \sin 3\omega t + \ldots \qquad [16]$$

To obtain the value of the waveform after half a cycle, i.e. at time $t + \pi$, we put this value of time into equation [16]:

$$v \text{ at } (t + \pi) = \tfrac{1}{2}a_0 + a_1 \cos 1\omega(t + \pi) + a_2 \cos 2\omega(t + \pi)$$
$$+ a_3 \cos 3\omega(t + \pi) + \ldots + b_1 \sin 1\omega(t + \pi)$$
$$+ b_2 \sin 2\omega(t + \pi) + b_3 \sin 3\omega(t + \pi) + \ldots$$

$$= \tfrac{1}{2}a_0 - a_1 \cos 1\omega t + a_2 \cos 2\omega t - a_3 \cos 3\omega t + \ldots$$
$$- b_1 \sin 1\omega t + b_2 \sin 2\omega t - b_3 \sin 3\omega t + \ldots \qquad [17]$$

If the waveform is to have negative half-cycles which are just the positive half-cycles inverted we must have the waveform after half a cycle, i.e. at time $t + \pi$, which is $-v$ at t. Thus we must have:

$$\tfrac{1}{2}a_0 + a_1 \cos 1\omega t + a_2 \cos 2\omega t + a_3 \cos 3\omega t + \ldots$$
$$+ b_1 \sin 1\omega t + b_2 \sin 2\omega t + b_3 \sin 3\omega t + \ldots$$
$$= - (\tfrac{1}{2}a_0 - a_1 \cos 1\omega t + a_2 \cos 2\omega t - a_3 \cos 3\omega t + \ldots$$
$$- b_1 \sin 1\omega t + b_2 \sin 2\omega t - b_3 \sin 3\omega t + \ldots) \qquad [18]$$

This can only occur if $a_0 = 0$, $a_2 = 0$, and all even harmonics are 0.

If the waveform is to have waveforms which repeat themselves after half a cycle then we must have the waveform at time $t + \pi$ equal to v at time t. Thus we must have:

$$\tfrac{1}{2}a_0 + a_1 \cos 1\omega t + a_2 \cos 2\omega t + a_3 \cos 3\omega t + \ldots$$
$$+ b_1 \sin 1\omega t + b_2 \sin 2\omega t + b_3 \sin 3\omega t + \ldots$$
$$= \tfrac{1}{2}a_0 - a_1 \cos 1\omega t + a_2 \cos 2\omega t - a_3 \cos 3\omega t + \ldots$$
$$- b_1 \sin 1\omega t + b_2 \sin 2\omega t - b_3 \sin 3\omega t + \ldots \qquad [19]$$

This can only occur if $a_1 = 0$, $a_3 = 0$ and all odd harmonics are 0.

Revision

3   Determine the nature of the terms within the Fourier series for the waveforms shown in Figure 11.7. T is the periodic time for a cycle.

11.3 Circuit analysis with complex waveforms

Often in considering electrical systems the input is not a simple d.c. or sinusoidal a.c. signal but perhaps a square wave periodic signal or a distorted sinusoidal signal or a half-wave rectified sinusoid. Such problems can be tackled by representing the waveform as a Fourier series and using the *principle of superposition*; we find the overall effect of the waveform by summing the effects due to each term in the Fourier series considered alone. Thus if we have a voltage waveform:

$$v = V_0 + V_1 \sin \omega t + V_2 \sin 2\omega t + V_3 \sin 3\omega t + \ldots \qquad [20]$$

then we can consider the effects of each element taken alone. Thus we can calculate the current due to the voltage V_0, that due to $V_1 \sin \omega t$, that due to $V_2 \sin 2\omega t$, that due to $V_3 \sin 3\omega t$, and so on for all the terms in the series. We then add these currents to obtain the overall current due to the waveform.

11.3.1 Pure resistance

Consider the application to a pure resistance R of a voltage of:

$$v = V_0 + V_1 \sin \omega t + V_2 \sin 2\omega t + \ldots + V_n \sin n\omega t \qquad [21]$$

where n is the number of the harmonic. Since $i = v/R$ and resistance R is independent of frequency, then the current due to the V_0 term is V_0/R, that due to the first harmonic term is $(V_1 \sin \omega t)/R$, that due to the second harmonic term is $(V_2 \sin 2\omega t)/R$ and so on. Thus the resulting current waveform is:

$$i = \frac{V_0}{R} + \frac{V_1}{R} \sin \omega t + \frac{V_2}{R} \sin 2\omega t + \ldots + \frac{V_n}{R} \sin n\omega t \qquad [22]$$

Because the resistance is the same for each harmonic, the amplitude of each voltage harmonic is reduced by the same factor, i.e. the resistance. The phases of each harmonic are not changed. The current waveform is thus the same shape as the voltage waveform.

Example

A complex voltage of $2.5 + 3.2 \sin 100t + 1.6 \sin 200t$ V is applied across a resistor having a resistance of 100 Ω. Determine the current through the resistor.

The complex current will be the sum of the currents due to each of the voltage terms in the complex voltage. Since the resistance is the same at all frequencies, the complex current will be:

$$i = 0.025 + 0.032 \sin 100t + 0.016 \sin 200t \text{ A}$$

Revision

4   Determine the waveform of the current occurring when a resistor of resistance 1 kΩ has connected across it the half-wave rectified sinusoidal voltage $v = 0.32 + 0.5 \cos 100t + 0.21 \cos 200t$ V.

11.3.2 Pure inductance

Consider the application to a pure inductance L of a voltage of:

$$v = V_0 + V_1 \sin \omega t + V_2 \sin 2\omega t + \; ... \; + V_n \sin n\omega t \tag{23}$$

where n is the number of the harmonic. The impedance of a pure inductance depends on the frequency, i.e. its reactance $X_L = \omega L$. Also the current lags the voltage by 90°. The impedance is 0 when the frequency is 0 and thus the current due to the V_0 term will be 0. The current due to the first harmonic will be the voltage of that harmonic divided by the impedance at that frequency and so $V_1 \sin (\omega t - 90°)/\omega L$. The current due to the second harmonic will be the voltage of that harmonic divided by the impedance at that frequency and so $V_1 \sin (2\omega t - 90°)/2\omega L$. Thus the current waveform will be

$$i = \frac{V_1}{\omega L} \sin(\omega L - 90°) + \frac{V_2}{2\omega L} \sin(2\omega L - 90°) + ...$$

$$+ \frac{V_n}{n\omega L} \sin(n\omega L - 90°) \tag{24}$$

Each of the voltage terms has its amplitude altered by a different amount; the phase, however, is changed by the same amount. The result is that the shape of the current waveform is different to that of the voltage waveform.

In terms of phasors, since the impedance $Z = \mathbf{V}/\mathbf{I}$, then for each of the Fourier series terms we have:

$$Z_n = \frac{\mathbf{V_n}}{\mathbf{I_n}} = \frac{V_n \angle \pi/2}{I_n} = n\omega L \angle 90° \tag{25}$$

$n\omega L$ is the reactance X_L for the harmonic concerned. We can represent the above relationship in complex notation as $Z_n = jX_L = jn\omega L$.

Example

A complex voltage of $2.5 + 3.2 \sin 100t + 1.6 \sin 200t$ V is applied across a pure inductor having a inductance of 100 mH. Determine the current through the inductor.

The impedance is 0 when the frequency is 0 and thus the current due to the 2.5 V term will be 0. For the second term, the reactance is 100 × 0.100 = 10 Ω and the current lags the voltage by 90° and so the current due to this harmonic is 0.32 sin (100t – 90°) A. For the third term, the reactance is $200 \times 0.100 = 20$ Ω and the current lags the voltage by 90° and so the current due to this harmonic is 0.08 sin $(100t - 90°)$ A. Thus the current waveform is:

$$i = 0.32 \sin (100t - 90°) + 0.08 \sin (100t - 90°) \text{ A}$$

Revision

5   Determine the waveform of the current when a pure inductor of inductance 10 mH has connected across it the half-wave rectified sinusoidal voltage $v = 0.32 + 0.5 \cos 100t + 0.21 \cos 200t$ V.

11.3.3 Pure capacitance

Consider a pure capacitor capacitance C when the voltage applied across it is:

$$v = V_0 + V_1 \sin \omega t + V_2 \sin 2\omega t + \ldots + V_n \sin n\omega t \qquad [26]$$

with n being the number of the harmonic. The impedance of a pure capacitor depends on the frequency, i.e. its reactance $X_C = 1/\omega C$, and the current leads the voltage by 90°. The impedance is 0 when the frequency is 0 and thus the current due to the V_0 term will be 0. The current due to the first harmonic will be the voltage of that harmonic divided by the impedance at that frequency and so $V_1 \sin (\omega t + 90°)/(1/\omega C)$. For the second harmonic the current will be the voltage of that harmonic divided by the impedance at that frequency and so $V_1 \sin (2\omega t + 90°)/(1/2\omega C)$. Thus the current waveform will be:

$$i = \omega C V_1 \sin(\omega t + 90°) + 2\omega C V_2 \sin(2\omega t + 90°) + \ldots$$
$$+ n\omega C V_n \sin(n\omega t + 90°) \qquad [27]$$

Each of the voltage terms has had their amplitude altered by a different amount but the phase changed by the same amount. The result of this is that the shape of the current waveform is different to that of the voltage waveform.

In terms of phasors, since the impedance $Z = \mathbf{V}/\mathbf{I}$, then for each of the Fourier series terms we have:

$$Z_n = \frac{\mathbf{V}_n}{\mathbf{I}_n} = \frac{V_n}{I_n \angle \pi/2} = \frac{1}{n\omega C} \angle (-90°) \qquad [28]$$

$1/n\omega C$ is the reactance X_C for the harmonic concerned. We can represent the above relationship in complex notation as $Z_n = -jX_C = -j(1/n\omega C)$ or $1/jn\omega C$.

Example

Determine the waveform of the current occurring when a 2μF capacitor has connected across it the half-wave rectified sinusoidal voltage $v = 0.32 + 0.5 \cos 100t + 0.21 \cos 200t$ V.

There will be no current arising from the d.c. term. For the first harmonic the reactance is $1/(2 \times 10^{-6} \times 100)$ Ω and so we have a current of $0.5 \times 2 \times 10^{-6} \times 100 \cos (100t + 90°)$ A. For the second harmonic the reactance is $1/(2 \times 10^{-6} \times 200)$ Ω and so the current is $0.21 \times 2 \times 10^{-6} \times 200 \cos (200t + 90°)$. Thus the resulting current is:

$$i = 2 \times 10^{-6} \times 0.5 \times 100 \cos (100t + 90°)$$
$$+ 2 \times 10^{-6} \times 0.21 \times 200 \cos (200t + 90°) \text{ A}$$

Revision

6   A voltage of $2.5 + 3.2 \sin 100t + 1.6 \sin 200t$ V is applied across a 10 μF capacitor. Determine the current.

11.3.4 Circuit elements in series or parallel

For circuit elements in series, the total impedance is the sum of the impedances of the separate elements. Thus if we have an inductance L in series with resistance R when there is an input of a voltage having harmonics, the impedance Z_n of the nth harmonic is the sum of the impedances for the nth harmonic of the two elements and is thus:

$$Z_n = R + jn\omega L \qquad [29]$$

If we had a resistance R, an inductance L and a capacitance C in series, then:

$$Z_n = R + jn\omega L + \frac{1}{jn\omega C} \qquad [30]$$

For parallel circuits, say a resistance R in parallel with an inductance L, when there is an input of a voltage having harmonics, the total impedance Z_n of the nth harmonic is given by:

$$\frac{1}{Z_n} = \frac{1}{R} + \frac{1}{jn\omega L} \qquad [31]$$

If we had a resistance R, an inductance L and a capacitance C in parallel then the total impedance for the nth harmonic is given by:

$$\frac{1}{Z_n} = \frac{1}{R} + \frac{1}{jn\omega L} + \frac{1}{1/jn\omega C} \qquad [32]$$

Example

A voltage of $v = 100 \cos 314t + 50 \sin(5 \times 314t - 30°)$ V is applied to a series circuit consisting of a 10 Ω resistor, a 0.02 H inductor and a 50 μF capacitor. Determine the circuit current.

For the first harmonic, the resistance is 10 Ω, the inductive reactance is $\omega L = 314 \times 0.02 = 6.28$ Ω and the capacitive reactance is $1/\omega C = 1/(314 \times 50 \times 10^{-6}) = 63.69$ Ω. Thus the total impedance for the first harmonic is:

$$Z_1 = 10 + j6.28 - j63.69 = 10 - j57.41$$

$$= \sqrt{10^2 + 57.41^2} \angle \tan^{-1} \frac{-57.41}{10} = 58.3 \angle (-80.1°)$$

Thus the current due to the first harmonic is:

$$i_1 = \frac{100 \angle 0°}{58.3 \angle (-80.1°)} = 1.72 \angle 80.1°$$

For the fifth harmonic, the resistance is 10 Ω, the inductive reactance is $5\omega L = 5 \times 314 \times 0.02 = 31.4$ Ω and the capacitive reactance is $1/5\omega C = 1/(5 \times 314 \times 50 \times 10^{-6}) = 12.74$ Ω. Thus the total impedance is:

$$Z_5 = 10 + j31.4 - j12.74 = 10 + j18.66$$

$$= \sqrt{10^2 + 18.66^2} \angle \tan^{-1} \frac{18.66}{10} = 21.2 \angle 61.8°$$

Thus the current due to the fifth harmonic is:

$$i_5 = \frac{50 \angle (-30°)}{21.2 \angle 61.8°} = 2.36 \angle (-91.8°)$$

Thus the current waveform is:

$$i = 1.72 \cos(314t + 80.1°) + 2.36 \cos(3 \times 314t - 91.8°) \text{ A}$$

Revision

7   A voltage of $200 \cos 314t - 40 \sin 2 \times 314t$ V is applied to a circuit consisting of a 20 Ω resistor in series with a 100 μF capacitor. Determine the current in the circuit.

8   A voltage of $1.0 \sin 500t + 0.2 \sin 3 \times 500t$ V is applied to a circuit consisting of three series components, a resistance of 500 Ω, an inductance of 1 H and a capacitance of 1 μF. Determine the current.

9   A triangular voltage waveform given by the Fourier series $v = 123 - 100 \cos 400t - 11 \cos 1200t - 4 \cos 2000t$ V is applied to a circuit consisting of three series components, a resistance of 40 Ω, an

inductance of 0.1 H and a capacitance of 25 μF. Determine the circuit current.

11.3.5 Selective resonance

The presence of harmonics in a waveform may give rise to *selective resonance*. This is when a circuit containing both inductance and capacitance resonates at one of the harmonic frequencies. It thus occurs, for a series circuit or parallel circuit involving inductance and capacitance, when the inductive reactance of a particular harmonic is equal to the capacitance reactance of that harmonic, i.e. when:

$$n\omega L = \frac{1}{n\omega C} \qquad\qquad [33]$$

where n is the number of the harmonic. When selective resonance occurs, the magnitude of the harmonic concerned is greatly increased. This has the effect of distorting the current waveform. It can also give rise to dangerously high voltage drops across the inductance and capacitance in the circuit. Selective resonance can also be beneficially used for tuning a circuit to a particular harmonic if that harmonic is to be enhanced while the others are reduced.

Example

The voltage $300 \sin \omega t + 100 \sin 3\omega t$ V is applied to a series circuit consisting of resistance 10 Ω, inductance 0.5 H and capacitance 0.2 μF. Determine (a) the fundamental frequency for resonance with the third harmonic, and (b) the current waveform at that frequency.

(a) Resonance with the third harmonic occurs when (equation 33) $n\omega L = 1/n\omega C$ and so:

$$\omega = \frac{1}{n}\sqrt{\frac{1}{LC}} = \frac{1}{3}\sqrt{\frac{1}{0.5 \times 0.2 \times 10^{-6}}} = 1054 \text{ rad/s}$$

and hence $f = \omega/2\pi = 168$ Hz.
(b) The impedance of the circuit at the fundamental is:

$$Z_1 = R + j\left(\omega L - \frac{1}{\omega C}\right)$$

$$= 10 + j\left(1054 \times 0.5 - \frac{1}{1054 \times 0.2 \times 10^{-6}}\right)$$

$$= 10 - j4217 = 4217\angle(-89.9°)\ \Omega$$

Thus the current at the fundamental is $300\angle 0°/[4217\angle(-89.9°)] = 0.071\angle 89.9°$ A. The impedance at the third harmonic is:

$$Z_3 = R + j\left(3\omega L - \frac{1}{3\omega C}\right)$$

But, because of selective resonance, $3\omega L = 1/3\omega C$ and so $Z_3 = R$. Thus the current at the third harmonic is $100\angle 0°/10 = 10\angle 0°$ A. The resulting current waveform is thus:

$$i = 0.071 \sin (1054t + 89.9°) + 10 \sin 3 \times 1054t$$

Revision

10 The voltage $10 \sin \omega t + 2 \sin 3\omega t$ V is applied to a series circuit consisting of resistance 5 Ω, inductance 0.1 H and capacitance 0.1 μF. Determine (a) the fundamental frequency for resonance with the third harmonic, and (b) the current waveform at that frequency.

11.4 Production of harmonics

The waveforms produced by a.c. generators is usually very nearly perfectly sinusoidal and thus the amount of harmonics is small. Most of the harmonic content of waveforms in circuits is, however, produced by non-linear circuit elements. A *non-linear circuit element* is one for which the current flowing through it is not proportional to the voltage across it. Examples of such devices are semiconductor diodes, transistors and ferromagnetic-cored coils. As an illustration, consider a transistor with the relationship between current and voltage shown in Figure 11.8. A sinusoidal voltage input gives rise to a distorted sinusoidal current.

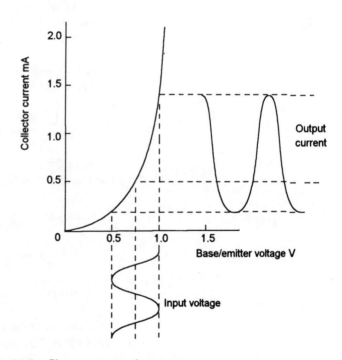

Figure 11.8 *Characteristic of a transistor*

Another example of non-linearity is that given by an iron-cored circuit element such as a transformer. When a sinusoidal alternating voltage is applied to a coil wound on an iron core, the current through the coil produces magnetic flux in the core. The relationship between the current and the magnetic flux is not linear. Indeed, not only is it not linear but an increasing value of current produces a different flow density to the same value of current when it is decreasing, this effect being termed hysteresis. Figure 11.9 illustrates this and shows how a sinusoidal current produces magnetic flux which varies with time in a non-sinusoidal manner. Because the core flux is non-sinusoidal, the e.m.f. induced in the secondary winding of a transformer by the magnetic flux will be non-sinusoidal.

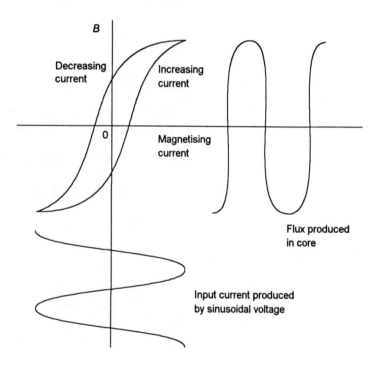

Figure 11.9   *Effect of iron core on magnetic flux waveform*

The current–voltage relationship for non-linear circuit elements can, in general, be represented by an equation of the form:

$$i = a + bv + cv^2 + dv^3 + \dots \qquad [34]$$

where a, b, c, d, etc. are constants. Suppose we have an applied voltage of $v = V \sin \omega t$, then if we consider just the a, b, c terms:

$$i = a + bV \sin \omega t + c(V \sin \omega t)^2$$

But $\sin^2 \omega t = \frac{1}{2}(1 - \cos 2\omega t)$ and so:

$$i = a + bV \sin \omega t + \frac{1}{2}cV^2(1 - \cos 2\omega t) \qquad [35]$$

Thus a second harmonic has been introduced. The amount of distortion resulting from this can be expressed as the amplitude of the second harmonic as a percentage of the fundamental.

$$\% \text{ second harmonic} = \frac{\frac{1}{2}cV^2}{bV} \times 100\% = \frac{cV}{2b} \times 100\% \qquad [36]$$

A typical amplifier stage response has a current–voltage characteristic of the form:

$$i = a + bv + cv^2 \qquad [37]$$

The input v is generally made up of a bias voltage V_b plus the signal to be amplified. If this is a sinusoidal signal $V_s \sin \omega t$ then equation [37] gives:

$$i = a + b(V_b + V_s \sin \omega t) + c(V_b + V_s \sin \omega t)^2$$

$$= a + bV_b + bV_s \sin \omega t + cV_b^2 + V_s^2 \sin^2 \omega t + 2cV_bV_s \sin \omega t$$

$$= a + bV_b + cV_b^2 + (bV_s + 2cV_bV_s) \sin \omega t + cV_s^2 \sin^2\omega t$$

$$= a + bV_b + cV_b^2 + (bV_s + 2cV_bV_s) \sin \omega t + \frac{1}{2}cV_s^2(1 - \cos 2\omega t)$$

When the alternating signal is zero then $i = a + bV_b + cV_b^2$. Let this current be represented by d. If we represent $(bV_s + 2cV_bV_s)$ by e and $\frac{1}{2}cV_s^2$ by f then:

$$i = d + e \sin \omega t + f - f \cos 2\omega t \qquad [38]$$

$d + f$ is the steady direct current, e is the peak value of the fundamental and f the peak value of the second harmonic. The percentage second harmonic distortion is thus $(f/e) \times 100\%$. We can write this in terms of current values occurring with the current waveform. The alternating current output will alternate between a maximum value, which occurs when $\omega t = 90°$, of $i_{max} = d + e + f + f$ and a minimum value, which occurs when $\omega t = 270°$, of $i_{min} = d - e + f + f$. Thus $i_{max} + i_{min} = 2d + 4f$. But d is the current occurring when the signal has a zero value. If we designate this as i_0 then $i_{max} + i_{min} - 2i_0 = 4f$. The maximum current value minus the minimum current value is $(d + e + f + f) - (d - e + f + f) = 2e$. Hence:

$$\% \text{ 2nd harmonic} = \frac{f}{e} \times 100\% = \frac{i_{max} + i_{min} - 2i_0}{2(i_{max} - i_{min})} \times 100\% \quad [39]$$

Example

A non-linear circuit element gave the current–voltage characteristic $i = 2.0 + 1.2v + 0.6v^2$ mA. Determine the percentage second harmonic content when $v = 1.0 \sin \omega t$ V.

Using $\sin^2 \omega t = \frac{1}{2}(1 - \cos 2\omega t)$:

$$i = 2.0 + 1.2 \times 1.0 \sin \omega t + \frac{1}{2} \times 0.6 \times 1.0^2(1 - \cos 2\omega t)$$

$$= 2.3 + 1.2 \sin \omega t - 0.3 \cos \omega t$$

The percentage second harmonic content is $(0.3/1.2) \times 100 = 25\%$.

Example

The current waveform produced by an amplifier stage is found to alternate between 60 mA and 85 mA when there is an input of a sinusoidal voltage with a constant current of 72 mA being given when the signal voltage is zero. If the amplifier characteristic is of the form $i = a + bv + cv^2$, determine the percentage second harmonic content.

Using equation [39]:

$$\% \text{ 2nd harmonic} = \frac{i_{max} + i_{min} - 2i_0}{2(i_{max} - i_{min})} \times 100\%$$

$$= \frac{85 + 60 - 2 \times 72}{2(85 - 72)} \times 100 = 3.8\%$$

Revision

11 A non-linear circuit element gave the current–voltage characteristic $i = 2.6 + 1.0v + 0.3v^2$ mA. Determine the percentage second harmonic content when $v = 1.0 \sin \omega t$ V.

12 A transistor amplifier stage has the current–voltage characteristic $i = 0.5 - 20v + 200v^2$ mA when v is in volts. Determine (a) the steady direct current and (b) the percentage second harmonic content when $v = 0.65 + 0.05 \sin \omega t$ V.

11.4.1 Circuits with rectified a.c.

Consider the effect of using a rectifier, or rectifiers, on a sinusoidal current. The result with half-wave rectification is of the form shown in Figure 11.10(a) and with full-wave rectification as shown in Figure 11.10(b). With the half-wave rectification the output can be expressed as:

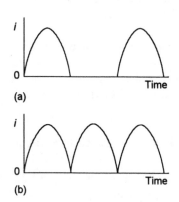

Figure 11.10 (a) Half-wave rectification, (b) full-wave rectification

$$i = \frac{A}{\pi}\left[1 + \frac{\pi}{2}\sin\omega t - 2\left(\frac{1}{3}\cos 2\omega t - \frac{1}{15}\cos 4\omega t + ...\right)\right]$$ [40]

and with full-wave rectification as:

$$i = \frac{2A}{\pi}\left[1 - 2\left(\frac{1}{3}\cos 2\omega t - \frac{1}{15}\cos 4\omega t + ...\right)\right]$$ [41]

We can thus carry out the analysis of circuits involving such signals in terms of their Fourier series.

Problems

1   Plot the graph of the waveform $v = 3\sin\omega t + \sin 2\omega t$.

2   Plot the graph of the waveform $v = 2\sin\omega t + \cos 3\omega t$.

3   Determine the nature of the terms within the Fourier series for the waveforms shown in Figure 11.11. T is the periodic time for a cycle.

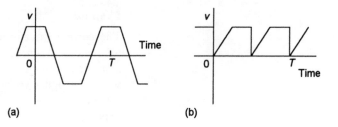

(a)   (b)

Figure 11.11   *Problem 3*

4   A voltage of $2\sin 500t + 1\sin 1000t$ V is applied across a resistor having a resistance of 10 Ω. Determine the circuit current.

5   A voltage of $2\sin 500t + 1\sin 1000t$ V is applied to a circuit consisting of a resistor of 6 Ω in series with a capacitor having a reactance of 8 Ω at $\omega = 500$ rad/s. Determine the circuit current.

6   A voltage of $50\sin\omega t + 25\sin(3\omega t + 60°)$ V is applied to a series circuit consisting of a resistor of 8 Ω, an inductor having a reactance of 2 Ω at the fundamental frequency and a capacitor having a reactance of 8 Ω at the fundamental frequency. Determine the circuit current.

7   A voltage of $10\sin 500t + 5\sin 1500t$ V is applied to a circuit consisting of a resistor of 5 Ω in parallel with an inductor having a reactance of 5 Ω at $\omega = 500$ rad/s. Determine the total circuit current.

8   A voltage of $25 + 100\sin 10\,000t + 40\sin(30\,000t + 30°)$ V is applied to a circuit consisting of a resistor of 5 Ω in series with an inductor of 0.5 mH. Determine the circuit current.

9   The voltage $25 \sin \omega t + 2 \sin 3\omega t$ V is applied to a series circuit consisting of a resistance of 5 Ω, an inductance of 0.4 H and a capacitance of 0.5 μF. Determine (a) the fundamental frequency for resonance with the third harmonic, and (b) the current waveform at that frequency.

10   A series circuit consists of a resistance of 20 Ω, an inductance of 10 mH and a capacitance of 0.1 μF. Determine the fundamental frequency if the circuit selectively resonates at the fifth harmonic.

11   A non-linear circuit element gave the current–voltage characteristic $i = 2.6 + 1.4v + 0.6v^2$ mA. Determine the percentage second harmonic content when $v = 2.0 \sin \omega t$ V.

12   The current waveform produced by an amplifier stage is found to alternate between 2 mA and 12 mA when there is an input of a sinusoidal voltage with a constant current of 6 mA being given when the signal voltage is zero. If the amplifier characteristic is of the form $i = a + bv + cv^2$, determine the percentage second harmonic content.

12 Systems

12.1 Introduction

The systems approach to engineering involves attention being focused on the functions of elements rather than how an effect is achieved. In these days of integrated circuits the question is generally which circuit to use, based on a consideration of the function required, rather than a consideration of the circuitry involved in the design of the integrated circuit or the workings of the constituent elements of that circuit. Thus, for example, we might require an integrated circuit to take a number of inputs and give an output when all of them are high signals. This consideration of systems enables the same basic approach to be used for a wide variety of engineering processes. For example, a measurement system consists of three basic system elements: a sensor, signal conditioning and display. We can apply this model to a simple measurement system such as a Bourdon gauge used for pressure measurement where the entire system is 'mechanical' and involves a Bourdon tube which is used to rotate, via gearing, a pointer across a scale or to an electronic measurement system for pressure measurement involving a semiconductor pressure element with a microprocessor used to process the signal and give a display.

12.2 Basic principles

The term system can be defined as:

> A *system* is a set of components which are connected together to accomplish a useful task.

Thus an amplifier can be considered to be a system, consisting of components such as transistors and resistors connected together to accomplish the task of taking an input signal and making it bigger. All systems are considered in terms of their inputs and outputs and you can think of a system as being rather like a machine into which you feed an input, the machine then processes the input and spews out its output.

We can have a number of systems connected together to achieve some function. For example a hi-fi music system might involve a CD player, a cassette player, a record player, an amplifier and loudspeakers. It is convenient to think of the systems as being interconnected by means of their inputs and outputs. Thus, for the hi-fi system, the output of the CD player becomes the input to the amplifier, the output of the amplifier becomes the input to the loudspeakers. A convenient way of displaying such interconnections is in terms of *block diagrams*.

With block diagrams, a rectangular box is used to represent a system, and inputs and outputs represented by lines with arrows (Figure 12.1(a)). The system represented by the box operates on the input signal to produce the output signal. In drawing block diagrams we also use other elements. A summing junction is represented by a circle (Figure 12.1(b)) with one or more arrowed lines for inputs coming in and an arrowed line for an output going out, plus or minus signs being placed against the input arrow heads to indicate whether the signals have to be added or subtracted. A take-off point (Figure 12.1(c)) allows a signal to be tapped and used elsewhere, the assumption being made that the signal is not affected by introducing the take-off point.

As an illustration of a block diagram, consider a domestic central heating system. It consists of a thermostat element which has inputs of the required temperature and the actual temperature and an output of a signal representing the difference between the required temperature and the actual temperature. This difference signal is used to control the central heating boiler and switch it on or off. The output from the boiler is used to give an input of heat to the room and so control its temperature. The output signal from the room system element is thus a temperature signal. A take-off point is used to indicate that the temperature signal is fed back to a summing junction where it is summed with the required temperature signal to give the difference signal.

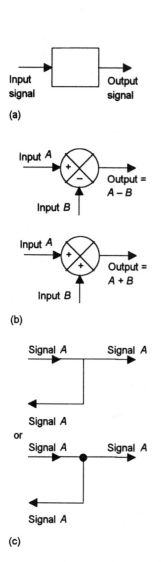

Figure 12.1   *Block diagram elements*

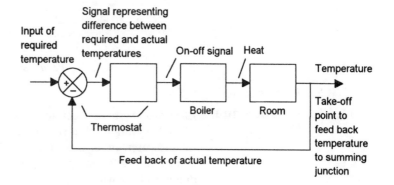

Figure 12.2   *Central heating system*

12.3 Mathematical models

A system takes an input, performs some action on it and then gives an output which is somehow related to the input. The output is thus some function of the input. In the case of an amplifier system we might have an input voltage V_{in} and an output voltage V_{out} which is proportional to the input and so:

$$V_{out} = A v_{in}$$
[1]

where A is the factor by which the amplifier multiplies the input, i.e. the amplifier gain. We can represent such a system by the block diagram

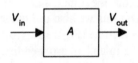

Figure 12.3   *Amplifier system*

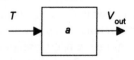

Figure 12.4   *Thermocouple system*

shown in Figure 12.3. The A in the box is the factor which operates on the input to give the output.

As another example, consider a sensor such as a thermocouple which has an input of temperature and an output of a voltage. When the voltage V_{out} is proportional to the temperature T we have (Figure 12.4):

$$V_{out} = aT \qquad [2]$$

with a being a constant and having the unit of V/°C.

12.3.1 Systems in series

A temperature measurement system might consist of a thermocouple connected to an amplifier and hence to a display (Figure 12.5). Suppose, for a steady temperature input, the thermocouple gives 4 mV/°C. When there is a temperature input to the system of, say, 10°C then the thermocouple gives an output of 40 mV. If the amplifier has a gain of 20 then the input to it of 40 mV becomes an output of 800 mV. The display might be a moving-coil voltmeter and give a pointer rotation of 1° per 50 mV. Thus the pointer rotates through 16° for the input to the system of 10°C.

Figure 12.5   *Temperature measurement system*

12.4 Information and signals

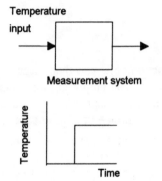

Figure 12.6   *Input to a measurement system*

Electronic systems react to *information*. This may take many forms. Thus there might be an electronic measurement system which is used for the measurement of temperature and so the system reacts to an input of temperature information. Another form of electronic system might be a digital camera which reacts to an input of the detailed information in the light coming from some scene to give a picture. Yet another form of electronic system is one used to count the number of items passing along a production line, the information then being in the form of on–off signals indicating the presence or not of items. As an illustration, Figure 12.6 shows the form the information might take for the temperature measurement system when the temperature sensor has its temperature suddenly increased. The input is then basically a step form of signal showing a sudden change in signal size.

The input signals to an electronic system will be converted by the system to time-varying voltages and currents which can then be processed by the system to give the system output. Between the input and output the signals can take a number of forms. Thus in a microprocessor-based system we might have the input signal from the sensor converted into a voltage signal which has an amplitude which

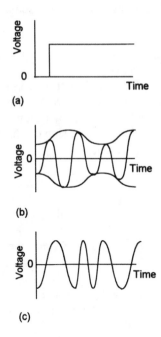

(a)

(b)

(c)

Figure 12.7   *Analogue signals*

varies with time, this then being converted into a digital signal where the pulse sequence is a measure of the amplitude variation, and then, after processing by the microprocessor, the digital output from the microprocessor is converted back into an output voltage which has an amplitude which varies with time.

12.4.1 Analogue and digital signals

With an *analogue* signal the information is carried by the size, amplitude or frequency of the signal varying with time. This might involve the size of the signal being changed, as in Figure 12.7(a). For an alternating signal the amplitude (Figure 12.7(b)) or frequency (Figure 12.7(c)) might be changed, the term used being modulated. With *amplitude modulation* (Figure 12.7(b)) the amplitude of the alternating carrier signal is varied according to how the information transmitted varies with time. With *frequency modulation* (Figure 12.7(c)), the modulating signal varies the frequency of the carrier signal as opposed to the amplitude in the case of amplitude modulation.

With a *digital* signal the information is conveyed as a number. This is generally by means of a signal which can assume just one of two levels, these levels representing the binary numbers 0 and 1. Thus, for example, information about an analogue signal of size 15 V is conveyed digitally by a signal representing the number 15 rather than a signal which is made a size which is related to 15 V. The numbers are conveyed in the binary system. The *binary system* is based on just two states 0 and 1, these being termed binary digits or *bits*. When a number is represented by this system, the digit position in the number indicates the weight attached to each digit with the weight increasing by a factor of 2 as we move from right to left in a number. The digit at the right-hand end is called the least significant bit (LSB) and the digit at the left-hand end the most significant bit (MSB). For example, the 4-bit binary number 1111 is:

$$\text{MSB} \underset{\displaystyle \underset{2^3}{\underset{2^2}{\overset{2^1}{2^0}}}}{1\ 1\ 1\ 1} \text{LSB}$$

and so is the decimal number $2^0 + 2^1 + 2^2 + 2^3 = 15$. The 8-bit binary number 10111011 is $2^0 + 2^1 + 0 + 2^3 + 2^4 + 2^5 + 0 + 2^7 = 187$. The combination of bits to represent a number is termed a *word*.

Digital signals can be conveyed either in parallel or serial. *Parallel transmission* with, say, an 8-bit word involves eight parallel wires with each bit of the word being transmitted along its own wire. *Serial transmission* involves the bits of the word being transmitted one after the other along a single wire.

12.5 Signal processing

Electronic circuits are used to process signals and the following are some of the types of processes involved.

12.5.1 Analogue to digital conversion

Signals from sensors are generally analogue and need to be converted to digital signals if the electronic system uses a microprocessor. An analogue-to-digital converter (ADC) has an input of an analogue signal and an output of a digital word which represent the analogue signal. The basic principles of analogue to digital conversion are that:

1   The analogue signal is sampled.

2   The sampled signal is then transformed into a digital signal.

3   Steps 1 and 2 are then repeated at regular intervals of time, the process being controlled by a clock.

A consequence of the sampling is that the ADC gives digital signals which represent the size of the analogue input signal at regular intervals of time rather than a continuous representation of the signal.

The relationship between the sampled analogue signal and the digital output is illustrated by Figure 12.8 where the output is in the form of a 3-bit word. With a 3 bit word there are just $2^3 = 8$ possible binary signals and so we divide the sampled analogue signal into 8 levels and represent each level by a word. The possible levels are called *quantisation levels* and the difference in analogue voltage between two adjacent levels is termed the *quantisation level*. Thus for the analogue to digital conversion represented by Figure 12.8, the quantisation interval is 1 V. When the analogue input is at a value which is centred over the quantisation interval the output word accurately represents the analogue signal. However, there will be an error when the analogue input is at other values, the maximum error being equal to one-half of the interval or ±½ bit. This is termed the *quantisation error*.

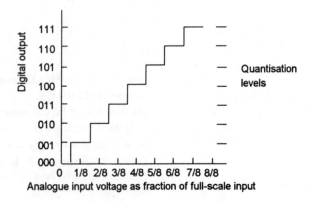

Figure 12.8 *Analogue to digital conversion*

The word length determines the *resolution*, i.e. the smallest changes in input which will result in a change in digital output, of the analogue-to-digital converter. The smallest change in digital output that can be registered is one bit in the least significant bit position in the word. Thus with a word length of n bits, since there are 2^n levels, the full-scale analogue input V_{FS} is divided into 2^n segments and so the smallest change in input that can be registered is $V_{FS}/2^n$. Thus:

$$\text{resolution} = \frac{V_{FS}}{2^n} \qquad [3]$$

Example

An analogue-to-digital converter has a word length of 8 bits and the analogue signal input varies between 0 and 10 V, what is the resolution?

The resolution is $10/2^8 = 0.039$ V. Any change less than this will fail to change the output. Thus an input of 0.030 V will give an output of 0000 0000, an input of 0.039 V an output of 0000 0001, an input of 0.060 V an output of 0000 0001, an input of 0.078 V an output of 0000 0010.

Revision

1   An electronic temperature measurement system employs a temperature sensor which gives an output which is converted to a digital signal by an analogue-to-digital converter and the digital signal is then displayed. If the range of temperatures to be measured is 0 to 200°C and a resolution of 0.5°C is required, what should be the word length of the ADC?

12.5.2 Digital to analogue conversion

The output from a microprocessor is digital and often needs to be converted into an analogue signal to operate devices such as control valves or motors. This is carried out by a digital-to-analogue converter (DAC). The input to a digital-to-analogue converter is a binary word and the output is an analogue signal that represents the word. Thus, for example, an input to a 3-bit DAC of 010 must give an analogue output which is twice that given by an input of 001 and an input of 011 must give an analogue output which is three times that of 001. Thus we might have 001 giving 1 V, 010 giving 2 V, 011 giving 3 V. Figure 12.9 illustrates this for an input to a 3-bit DAC for which each additional bit increases the output voltage by 1 V. The word length that can be handled by a DAC determines its resolution. Thus a 4-bit DAC with a full-scale voltage of V_{FS} gives an analogue output which changes in steps of $V_{FS}/2^4$ while an 8-bit DAC gives an output which changes in steps of $V_{FS}/2^8$.

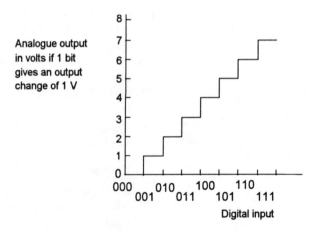

Analogue output
in volts if 1 bit
gives an output
change of 1 V

Figure 12.9   *Digital to analogue conversion*

Example

A microprocessor gives an output of an 8-bit word which is fed through an 8-bit digital-to-analogue converter to a control valve. The control valve requires the voltage to change from 0 to 10 V to go from fully closed to fully open. If the fully closed state is represented by 0000 0000 and the fully open state by 1111 1111 what will be the change in output to the valve when there is a change of 1 bit in the output from the microprocessor?

The full-scale output voltage of 10 V will be divided into 2^8 intervals. A change of 1 bit is thus a change in the output voltage of $10/2^8 = 0.039$ V.

12.5.3 Amplification

Amplifiers can be purchased as complete integrated circuits, the circuits containing many transistors and other components. The concern is then with the overall properties of an amplifier, i.e. the behaviour of the amplifier system, rather than how the circuit works.

Amplification is ideally the process of increasing in size or amplitude a signal without changing its waveform. We can consider the voltage gain, current gain or power gain. The *voltage gain* is the ratio of the input to output voltage, the *current gain* is the ratio of the input to output current and the *power gain* is the ratio of the input to output power.

We cannot, however, in practice assume that the gain applies to all input values or to all frequencies. The range within which the output signal can vary is invariably limited; this means that there is a maximum value of input that will produce an output signal for which the waveform is essentially unchanged from that of the input. Increasing the input signal beyond this value will produce *clipping* of the higher levels of the output (Figure 12.10). The gain of an amplifier does not remain constant

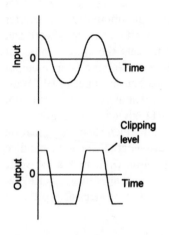

Figure 12.10   *Clipping*

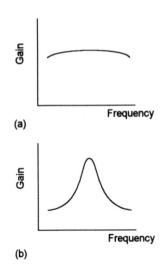

(a)

(b)

Figure 12.11   *Effect on gain of frequency: (a) direct coupled amplifier, (b) frequency selective amplifier*

with the frequency of the input signal. *Direct coupled amplifiers* have gains that remain essentially constant from zero frequency to a relatively high frequency (Figure 12.11(b)); *frequency selective amplifiers* (Figure 12.11(b)) are designed to amplify only over particular frequency ranges. The effect of the gain changing with frequency with a complex waveform is to produce different amplification of some of the harmonics that constitute it and so give an output waveform different from that of the input.

Gains are often expressed in decibels. The *bel* is a logarithmic unit of power ratio:

$$\text{power gain in bels} = \lg\left(\frac{\text{output power}}{\text{input power}}\right) \tag{4}$$

Because the bel is a rather large unit, decibels (dB) are used:

$$\text{power gain in dB} = 10\lg\left(\frac{\text{output power}}{\text{input power}}\right) \tag{5}$$

Decibels can also be used to express ratios of d.c. or a.c. voltages (r.m.s. or maximum values). Because the power dissipated in a resistance R when there is a voltage V across it is V^2/R it is convenient to assume that the input and output voltages are across the same resistance and so we can write equation [5] as:

$$\text{voltage gain in db} = 10\lg\left(\frac{\text{output voltage}^2}{\text{input voltage}^2}\right) \tag{6}$$

or

$$\text{voltage gain in db} = 20\lg\left(\frac{\text{output voltage}}{\text{input voltage}}\right) \tag{7}$$

Likewise, since the power dissipated in a resistance by a current I through it is I^2R:

$$\text{current gain in db} = 20\lg\left(\frac{\text{output current}}{\text{input current}}\right) \tag{8}$$

An amplifier might thus be specified as having a power gain of 15 dB.

An amplifier can be represented by the circuit shown in Figure 12.12. The input signal is connected across the amplifier input resistance R_{in} and the output signal is a voltage source, which has a voltage of the product of the amplifier voltage gain A_V and the input voltage V_{in}, in series with the amplifier output resistance R_{out}.

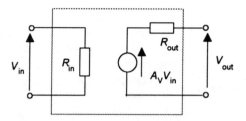

Figure 12.12   *Equivalent circuit for an amplifier*

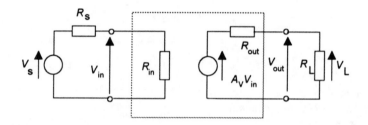

Figure 12.13   *An amplifier inserted between a source and load*

When in use, an amplifier is essentially connected to a voltage source V_s with an internal resistance R_s and with a load of resistance R_L across the amplifier output. (Figure 12.13). The two resistors R_{in} and R_s form a voltage divider so the input voltage to the amplifier is:

$$V_{in} = V_s \frac{R_{in}}{R_s + R_{in}} \qquad [9]$$

The larger the value of the amplifier input resistance the closer is the value of V_{in} to V_s. The voltage value obtained by multiplying the input voltage by the voltage gain of the amplifier, i.e. $A_V V_{in}$, is applied across the output resistance of the amplifier and the load resistance R_L. The voltage appearing across the load V_L is thus:

$$V_L = A_V V_{in} \frac{R_L}{R_L + R_{out}} \qquad [10]$$

Note that the $A_V V_{in}$ is the voltage produced when there is an open-circuit and is *not* the voltage produced across a load. The voltage gain A_V is thus the gain of an amplifier which is defined for the case of an open circuit load.

As an illustration, an amplifier might be specified as having a gain of 34 dB, an input impedance of 50 kΩ and an output impedance of 5 kΩ.

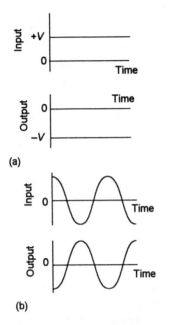

(a)

(b)

Figure 12.14 *Inversion with: (a) d.c. and (b) a.c. signals*

The following are some of the basic types of amplifiers that are encountered.

1 *Inverting and non-inverting amplifiers*
Amplification is often accompanied by inversion when a positive input gives rise to a negative output; Figure 12.14 illustrates this for d.c. and a.c. signals. Amplifiers giving inversion are termed inverting amplifiers and those giving no inversion are termed non-inverting amplifiers.

2 *Voltage and current amplifiers*
The term voltage amplifier is used for an amplifier designed to amplify the voltage of the input signal. Its voltage gain is specified but its current gain is not; it may give an increase in signal current if the load impedance is low enough. The term current amplifier is used for an amplifier designed to amplify the current of the input signal. Both types of amplifier may be inverting or non-inverting.

3 *Power amplifiers*
The term power amplifier is used where both the current and the voltage of the output are significant and the amplifier is required to supply signal power to a particular load. A typical specification might thus include: output power 40 W into a load of 8 Ω, output impedance less that 0.4 Ω, input resistance greater than 10 kΩ.

4 *Differential amplifiers*
A differential amplifier has two input signals and is required to amplify the difference between them.

5 *Buffer amplifiers*
A buffer amplifier is used as the interface between a high impedance signal source and a low impedance load. The gain is not significant and indeed is often 1.

Example

A sinusoidal voltage supply of 2 mV has an internal resistance of 9 kΩ and is connected to an amplifier of open-circuit voltage gain 100, input resistance 91 kΩ and output resistance 100 Ω. Determine the voltage appearing across a load of 1 kΩ connected across the output terminals of the amplifier.

The input voltage to the amplifier is given by equation [9] as:

$$V_{in} = V_s \frac{R_{in}}{R_s + R_{in}} = 2 \times \frac{91}{91+9} = 1.82 \text{ mV}$$

The voltage across the load is given by equation [10] as:

$$V_L = A_V V_{in} \frac{R_L}{R_L + R_{out}} = 100 \times 1.82 \times \frac{1000}{1000+100} = 165 \text{ mV}$$

Example

An amplifier is specified as having a voltage gain of 12 dB. Express this as a voltage ratio.

Using equation [7]:

$$12 = 20 \lg \text{(voltage ratio)}$$

We can write this as:

$$10^{12} = \text{(voltage ratio)}^{29}$$

and so

$$\text{voltage ratio} = 10^{12/20} = 3.98$$

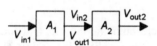

Figure 12.15   *Example*

Example

Two amplifiers with voltage gains of 12 dB and 5 dB are connected in cascade with the output of one feeding the input of the other. What is the overall voltage gain if it can be assumed that the second amplifier does not affect the gain of the first and vice versa?

For two systems connected in cascade (Figure 12.15) we have the output for the first system becoming the input of the second system. Thus if the gain of the first system is $A_1 = V_{out1}/V_{in1}$ then for the second system we have $A_2 = V_{out2}/V_{in2} = V_{out2}/V_{out1}$. The overall gain of the system is $V_{out2}/V_{in1} = (V_{out1}/V_{in1})(V_{out2}/V_{out1}) = A_1 \times A_2$. If we take logarithms then:

$$\lg \text{overall gain} = \lg A_1 + \lg A_2$$

and hence:

$$20 \lg \text{overall gain} = 20 \lg A_1 + 20 \lg A_2$$

Thus the overall gain in dB is equal to the sum of the gains of the two systems when expressed in dB. Hence the overall gain is 12 + 5 = 17 dB.

Example

Two amplifiers with voltage gains of 12 dB and 8 dB are connected in cascade by a cable of attenuation 3 dB with the output of one feeding the input of the other. What is the overall voltage gain if it can be assumed that the second amplifier does not affect the gain of the first and vice versa?

An attenuator is a network with a negative gain, in this case –3 dB. Thus we have three systems in cascade, the two amplifiers and the cable and so the overall gain is 12 – 3 + 8 = 17 dB.

Revision

2   A sinusoidal voltage supply of 10 mV has an internal resistance of 300 Ω and is connected to an amplifier with an open-circuit voltage gain of 100, input resistance 10 kΩ and output resistance 100 Ω. Determine the overall voltage gain and the power gain when the amplifier has a load resistance of 50 Ω connected across it.

3   An amplifier is specified as having a voltage gain of 18 dB. Express this as a voltage ratio.

4   Two amplifiers with voltage gains of 18 dB and 12 dB are connected in cascade with the output of one feeding the input of the other. What is the overall voltage gain if it can be assumed that the second amplifier does not affect the gain of the first and vice versa?

5   Two amplifiers with voltage gains of 10 dB and 8 dB are to be connected in cascade by cable of attenuation 1 dB/metre with the output of one feeding the input of the other. What is the overall voltage gain if it can be assumed that the second amplifier does not affect the gain of the first and vice versa and 1 m of cable is used?

12.5.4 Amplifiers with feedback

A feedback system is one for which the input signal to the system is modified in some way by the system output signal being fed back to the input. Consider the amplifier system shown in Figure 12.16 in which the output voltage is fed back so that it subtracts from the input voltage, the fed back voltage being in series with the input voltage. The term *negative feedback* is used.

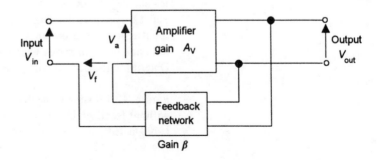

Figure 12.16 *Amplifier with feedback*

The input signal to the amplifier with negative feedback is:

$$V_a = V_{in} - V_f \hspace{5cm} [11]$$

The feedback signal is the output signal V_{out} after having passed through the feedback circuit. This circuit in its simplest form could be two resistors in series across the output so that the feedback signal is taken as the voltage across one of them, i.e. a voltage divider circuit. The result is that some proportion of the output signal is fed back. If this feedback circuit has a gain β then $V_f = \beta V_{out}$. Equation [11] can thus be written as:

$$V_a = V_{in} - \beta V_{out} \hspace{4cm} [12]$$

For the amplifier we have $V_{out} = A_V V_a$ and so:

$$V_{out} = A_V(V_{in} - \beta V_{out})$$

$$V_{out}(1 + \beta A_V) = A_V V_{in} \hspace{3.5cm} [13]$$

Thus the overall voltage gain with feedback is:

$$\text{overall gain} = \frac{V_{out}}{V_{in}} = \frac{A_V}{1 + \beta A_V} \hspace{2.5cm} [14]$$

The overall gain is less than A_V. If βA_V is much greater than 1 then the overall gain is effectively $1/\beta$ and so independent of the voltage gain of the basic amplifier.

The voltage gain of the basic amplifier without feedback depends on the load connected across its output and so is not stable. The voltage gain with feedback can, however, be obtained with a high degree of stability by selection of the component values of the feedback network.

Example

An amplifier has an open-circuit voltage gain of 1000 and an output resistance of 100 Ω. It is connected across a load of resistance 900 Ω. Negative feedback is provided by connection of a voltage divider across the output so that one tenth of the output voltage is fed back to be in series with the input voltage and subtract from it. Determine the overall voltage gain of the system.

Figure 12.17 shows the basic system. The gain of the amplifier without feedback is, when the loading effect of the feedback network is neglected:

$$\text{gain without feedback} = 1000 \times \frac{900}{100 \times 900} = 900$$

With negative feedback the gain, using equation [14], becomes:

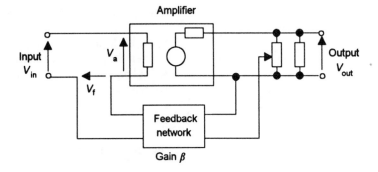

Figure 12.17 *Example*

$$\text{overall gain} = \frac{V_{out}}{V_{in}} = \frac{A_V}{1 + \beta A_V} = \frac{900}{1 + 0.1 \times 900} = 9.89$$

Example

An amplifier is required which will have an overall voltage gain of 100 and for which the gain will not vary by more than 1.0%. What gain and feedback attenuation will be required of a basic amplifier with negative feedback if load changes result in the basic amplifier voltage gain changing by as much as 20%?

With negative feedback the gain of the required amplifier is given by equation [14] as:

$$\text{overall gain} = 100 = \frac{A_V}{1 + \beta A_V}$$

$$A_V(1 - 100\beta) = 100$$

However, when the basic amplifier gain changes to $0.8A_V$ the overall gain must only change to 99. Thus:

$$99 = \frac{0.8A_V}{1 + \beta 0.8A_V}$$

$$A_V(0.8 - 79.2\beta) = 99$$

Eliminating A_V between the two equations gives:

$$99(1 - 100\beta) = 99(0.8 - 79.2\beta)$$

Hence $\beta = 0.0096$. Substituting this value in an equation gives $A_V = 2500$.

Revision

6   An amplifier has an open-circuit voltage gain of 100 and an output resistance of 100 Ω. It is connected across a load of resistance 900 Ω. Negative feedback is provided by connection of a voltage divider across the output so that one twentieth of the output voltage is fed back to be in series with the input voltage and subtract from it. Determine the overall voltage gain of the system.

12.5.5 Oscillators

The function of an oscillator is to produce a constant frequency, constant amplitude sinusoidal signal. Oscillators can be considered to be basically amplifiers with positive feedback (Figure 12.18). Equation [14] thus becomes:

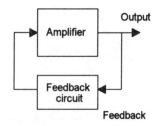

$$\text{overall gain} = \frac{V_{out}}{V_{in}} = \frac{A_V}{1 - \beta A_V} \qquad [15]$$

When $\beta A_V = 1$ then the overall gain becomes infinite; the system is thus unstable. This is the condition for oscillation.

If we think of the amplifier as initially having an input of a sinusoidal signal, then it gives an output of a larger sinusoidal signal. Some of this larger signal is then fed back to provide to the input. Provided it is in phase with the original input signal, i.e. the feedback is positive rather than negative, and the overall loop gain is 1 then the result is that the input has been 'replenished'. The circle can then repeat itself and so the system becomes self-perpetuating.

Figure 12.18   *Basic principle of an oscillator*

The conditions for sustained oscillations are:
1   The phase shift around the feedback loop must be 0°.
2   The voltage gain around the feedback loop must be 1.

The voltage gain around the closed feedback loop is the product of the amplifier gain A_V and the attenuation β of the feedback circuit, i.e. $A_V\beta$. The term *attenuation* is used for a network which has a negative gain, i.e. the output is smaller than the input. Thus since we must have $A_V\beta = 1$, if the amplifier has a voltage gain of 100 then the feedback circuit must have an attenuation of 0.01 to make the loop gain equal to 1.

In considering an oscillator the parameters that have to be taken into account are its frequency range, its frequency stability, its amplitude stability and its distortion level. There are two main forms of oscillator: *LC* oscillators which are generally used for the higher frequency ranges above about 10^5 Hz and *RC* oscillators which are used from very low frequencies up to about 10^7 Hz. The *LC* and *RC* refer to the type of network used for the feedback.

12.5.6 Noise

If the volume control of a hi-fi system is turned right up and there is no deliberate input signal to the system, it is often possible to hear a hiss from the loudspeakers. This arises from what is termed internal *noise*. Noise is the term used with electronic systems for unwanted signals that occur; they can be internally generated within a system or arise from external sources.

For internal noise, we can think of the electrons in a conductor as being like molecules in a gas and in random motion resulting from the effect of being at some temperature above absolute zero. This random motion leads to a randomly varying voltage appearing across a conductor; this is one source of internal noise. Another source of internal noise arises from a direct current being the average rate of flow of charge carriers with time and because the actual number may vary from second to second, a fluctuating signal is produced. Thus such noise can be generated by the electrons moving through a transistor. Internal noise can also arise from changes in the conductivity of semiconductor material. The contribution of each resistor and transistor to the noise output of an amplifier depends on its position in the circuit. Noise generated in the internal resistance of the signal source or the input of the amplifier is amplified by the amplifier and can become very significant and completely exceed any noise generated in a component at later stages in the amplifier. In addition to internally generated noise we can also have externally generated noise which perhaps can arise from electrostatic and electromagnetic interference from the alternating current mains supply.

For a particular signal input power over a defined frequency range, the signal-to-noise ratio for an amplifier is defined as:

$$\text{S/N ratio} = \frac{\text{wanted signal power}}{\text{noise signal power}} \qquad [16]$$

The ratio is usually expressed in decibels (dB) as:

$$\text{S/N ratio} = 10\lg\left(\frac{\text{wanted signal power}}{\text{noise signal power}}\right) \qquad [17]$$

The S/N ratio for a communication system determines the quality of the information received. An S/N ratio of about 70 dB gives a signal which appears to be essentially free from noise and is typical of the sound produced by a good quality hi-fi system. With an S/N ratio of 20 dB the noise becomes noticeable. This would be acceptable in a telephone system where speech is being transmitted but unacceptable with music. With an S/N ratio of 6 dB the signal is very badly degraded by noise and when the S/N ratio drops to −10 dB the signal becomes completely lost in noise.

With a digital system, the information processed is binary and the systems are only testing whether the signal received is high or low. The actual size of the high and low signals may vary quite significantly, the precise values not being important as long as the system can distinguish between high and low levels. Thus noise is no problem provided it is not large enough to induce a level change. Thus the transmission of data in digital form offers advantages over transmission in analogue form since an analogue signal will be degraded in transmission as a result of noise being picked up while the noise, provided it is not so large as to permit discrimination between high and low levels, will not affect the quality of transmission of a digital signal.

Example

If the S/N ratio for each of two cascaded amplifiers is 46 dB, what is the overall S/N ratio?

If the wanted signal power is taken as 1 W then the noise signal power for an amplifier is $10^{-4.6} = 2.5 \times 10^{-5}$ W. The total noise signal power is thus $2 \times 2.5 \times 10^{-5}$ W. Thus the total S/N ratio is 10 $\lg(1/5 \times 10^{-5}) = 43$ dB.

Revision

7   The S/N ratio for an amplifier is 40 dB. What is the noise power when there is a wanted signal power of 1 W?

12.6 Examples of electronic systems

As illustrations of the way the above electronics systems can be used, consider the following applications.

12.6.1 An AM receiver

Figure 12.19 shows a block diagram of the system used for an AM (amplitude modulation) receiver.

1   The aerial picks up transmitted radio signals. At this point the required frequency is just one of many that has been received.

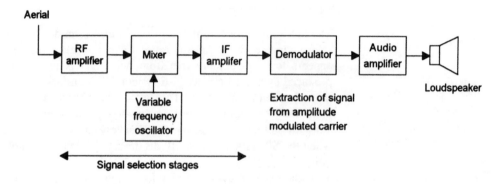

Figure 12.19   *An AM receiver*

2   The radio frequency (RF) amplifier amplifies the very small signals. This is a type of amplifier specifically designed for amplifying such high frequency signals.

3   The required carrier frequency is selected by first shifting the high frequency down to an intermediate (IF) frequency. This is achieved by mixing (heterodyning) it with the output from a variable frequency oscillator to give an output which is the sum and difference frequencies of the oscillator and the carrier frequency. The result is that the carrier frequencies can be shifted down to the standard intermediate frequency for AM receivers of about 460 Hz and tuning can be achieved by shifting just the required carrier frequency down to this frequency.

4   A further stage of amplification is then carried out, the IF amplifier being just needed to amplify signals at this 460 Hz frequency.

5   Demodulation is then used to extract the amplitude modulation signal from the carrier frequency.

6   An audio amplifier is then used to amplify the signals for conversion to sound by the loudspeakers.

12.6.2 Temperature measurement system

Consider the basic form of a temperature measurement system which involves the measurement of temperatures in the range 0 to 100°C to give a binary output with a change of 1 bit corresponding to a temperature change of 1°C. The output can then be fed to a microprocessor for further processing, perhaps as part of a temperature control system to give an output when the temperature reaches a particular value or perhaps a display which can be switched to indicate the actual temperature, the minimum temperature and the maximum temperature. Figure 12.20 shows the basic form such a system can take and Figure 12.21 a system that could be used. The following are the basic factors taken into account in the selection of the components in the system.

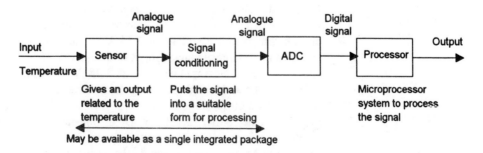

Figure 12.20   *Temperature measurement system*

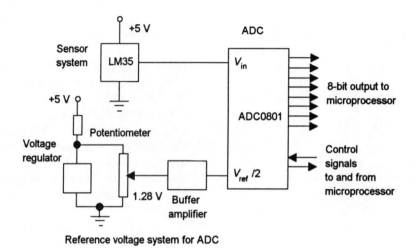

Reference voltage system for ADC

Figure 12.21　*Temperature measurement system*

1　A temperature sensor is required which gives an output proportional to the temperature so that each 1°C change can be used to generate 1 bit. A suitable sensor is the thermo-transistor LM35 (with a thermotransistor the voltage across the junction between the base and emitter depends on the temperature). LM35 gives an output of 10 mV/°C when it has a supply voltage of 5 V.

2　The output from the LM35, which is analogue, is then applied to an analogue-to-digital converter (ADC) to give a digital output. An 8-bit analogue-to-digital converter (ADC) is required since changes corresponding to a total range of 100 bits is required and we required $2^n = 100$ and so a 6.6 bit ADC; the nearest ADC is 8-bit. Since 1°C generates 10 mV, we need the resolution of the ADC to be 10 mV so that each step of 10 mV will generate a change in output of 1 bit.

3　Suppose we use an 8-bit successive approximations ADC, e.g. ADC0801. This type of ADC requires an input of a reference voltage which when subdivided into $2^8 = 256$ bits gives 10 mV per bit. Thus a reference voltage of 2.56 V is required. For this to be obtained the reference voltage input to the ADC0801 has to be $V_{ref}/2$ and so an accurate input voltage of 1.28 V is required. Such a voltage can be obtained by using a potentiometer circuit across a 5 V supply with a buffer amplifier (a voltage follower) to avoid loading problems (the buffer amplifier is a non-inverting operational amplifier where all the output voltage is fed back with no attenuation to an inverting input, the result being a total voltage gain of 1). The buffer amplifier has a very high input impedance and very low output impedance and is used for interfacing high impedance sources and low impedance loads. Because the voltage has to remain steady at 1.28 V, even if the 5 V supply voltage fluctuates, a voltage regulator is likely to be used, e.g. a 2.45 V voltage regulator ZN458/B.

4   Control signals to and from the microprocessor are used to ensure that the ADC supplies signals when the microprocessor is ready.

12.6.3 Humidity measurement system

Relative humidity is defined as being the mass of water vapour present in a volume of air when compared with the mass of water vapour required to saturate that volume at the same temperature. The mass of water vapour required to saturate a volume of air depends on the temperature.
The traditional method of measuring humidity involves two thermometers, one with its bulb directly exposed to the air and giving the 'dry temperature' and the other with its bulb covered with muslin which dips into water. The rate of evaporation from the wet muslin depends on the amount of water vapour present in the air; when the air is far from being saturated then the water evaporates quickly, when saturated there is not net evaporation. This rate of evaporation affects the temperature indicated by the thermometer, so giving the 'wet temperature'. Tables are then used to convert these readings into the humidity. Consider the design of a measurement system which will automatically display the humidity. Figure 12.22 shows the basic form a microprocessor-based system might take.

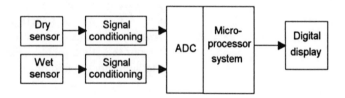

Figure 12.22   *Humidity measurement*

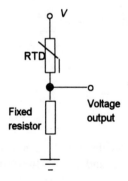

Figure 12.23   *Signal conditioning with an RTD*

The system components might be:

1   Temperature sensors such as the LM35 or resistance temperature detector (RTD), just a resistor which gives a resistance change with temperature change, with one exposed directly to the air and the other covered with a muslin sheath which dips in water.

2   Possibly signal conditioning to obtain a suitable voltage signal for analogue to digital conversion. In the case of an RTD this means converting the resistance change into a voltage. This can be achieved by the circuit shown in Figure 12.23, the RTD being in

series with a fixed resistor and connected across a constant voltage supply. The voltage drop across the fixed resistor depends on the total resistance of the circuit and hence on the resistance of the RTD.

3   Rather than use an ADC for each sensor, a single ADC can be used and the microprocessor programmed to sample first one sensor input and then the other. Microcontrollers are available with an ADC combined with a microprocessor in a single chip.

4   Traditionally the humidity was obtained by using a table and looking up the humidity indicated by the two temperatures. With a micro-processor system the table can be stored in its memory and the microprocessor 'looks up' the value.

5   The microprocessor then gives the result as a signal for the display.

A problem with the above system is that it requires a reservoir of water which has to be maintained in order that the wet sensor can function. An alternative which is now frequently used is a capacitive humidity sensor. The sensor (Figure 12.24) consists of an aluminium substrate with its top surface oxidised to form a porous layer of aluminium oxide. On top of the oxide a very thin gold layer is deposited, this being permeable to water vapour. Electrical connections are made to the gold layer and the aluminium substrate, the arrangement being a capacitor with an aluminium oxide dielectric. Water vapour enters the pores of the aluminium oxide and changes its dielectric constant and hence the capacitance of the capacitor. The capacitance thus gives a measure of the amount of water vapour present in the air. Figure 12.25 shows the type of system that might be used with such a sensor. A temperature sensor is also required since the maximum amount of water vapour that air can hold depends on the temperature and thus to compute the humidity the microprocessor needs to know the temperature.

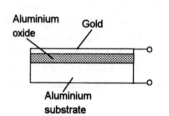

Figure 12.24   *Humidity sensor*

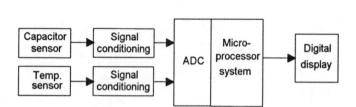

Figure 12.25   *Humidity measurement*

Problems   1   An analogue-to-digital converter has a word length of 12 bits and the analogue signal input varies between 0 and 5 V, what is the resolution?

2 An electronic gauge pressure measurement system employs a sensor which gives an output which is converted to a digital signal by an analogue-to-digital converter and the digital signal then displayed. If the range of gauge pressure to be measured is 0 to 100 kPa and a resolution of 0.5 kPa is required, what should be the word length of the ADC?

3 A microprocessor gives an output of a 4-bit word which is fed through an 4-bit digital-to-analogue converter to control the speed of a motor. The motor requires the voltage to change from 0 to 10 V to go from stopped to full speed. If the stopped state is given by 0000 and the full speed state by 1111 what will be the change in output to the motor when there is a change of 1 bit in the output from the microprocessor?

4 A sinusoidal voltage supply of 2 mV has an internal resistance of 1 kΩ and is connected to an amplifier of open-circuit voltage gain 10, input resistance 99 kΩ and output resistance 100 Ω. Determine the voltage appearing across a load of 1 kΩ connected across the output terminals of the amplifier.

5 An amplifier is specified as having a voltage gain of 26 dB. Express this as a voltage ratio.

6 Two amplifiers with voltage gains of 10 dB and 5 dB are connected in cascade with the output of one feeding the input of the other. What is the overall voltage gain if it can be assumed that the second amplifier does not affect the gain of the first and vice versa?

7 Two amplifiers with voltage gains of 12 dB and 10 dB are connected in cascade by a cable of attenuation 3 dB with the output of one feeding the input of the other. What is the overall voltage gain if it can be assumed that the second amplifier does not affect the gain of the first and vice versa?

8 A basic amplifier has voltage gain of 1000 and is to be used with negative feedback with an attenuation of 0.001. Determine the overall voltage gain of the feedback system.

9 An amplifier is required which will have an overall voltage gain of 100 and is to be constructed from a basic amplifier with a gain of 500 and negative feedback. Determine the attenuation required of the feedback.

10 The S/N ratio for each of three cascaded amplifiers is 46 dB, what is the overall S/N ratio?

13 Control systems

13.1 Introduction

Control systems are everywhere. They are used to control the speed of rotation of a d.c. or a.c. motor. They are used with the domestic central heating system so that the temperature in the house adjusts to maintain a set value. They are used with an automatic camera to adjust the lens position so that the object being photographed is always in focus and to ensure the exposure is correct. They are used with the domestic washing machine to ensure that each of the steps in the washing sequence are carried out in the right order. These are just a few examples of control systems. This chapter is a consideration of the basic principles of control systems and the elements involved in such systems.

13.2 Basic principles

The system used to control the speed of rotation of a motor might be open-loop or closed-loop. With an *open-loop system*, the motor speed is set by selecting the position of some control knob, the motor then runs to a speed indicated by the knob. However, if the load on the motor changes, there is no mechanism to adjust the motor to compensate for the load change and the speed drops. With a *closed-loop system*, the motor speed is set by the control knob and now, when the load changes there is a feedback signal from some output sensor back to the input to the motor to indicate that the change has occurred and, as a consequence, the speed is adjusted. The closed-loop system can thus take account of load changes, the open-loop system cannot. The primary advantage of open-loop control is that it is less expensive than closed-loop control. The disadvantage is that errors caused by disturbances or changes in loading are not corrected.

An *open-loop control system* does not compare the actual output with the required output to determine the control action but uses a calibrated input setting to obtain the required output. A *closed-loop system* uses feedback to compare the actual output with the required output and so modify its control action in the event of any difference to obtain the required output.

13.2.1 Elements in control systems

The elements of a basic open-loop and a basic closed-loop system are shown in Figure 13.1. The elements of the open-loop system are:

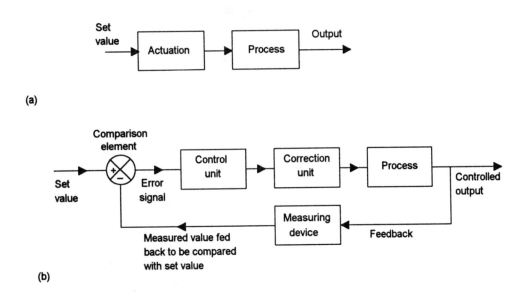

(a)

(b)

Figure 13.1 *Basic elements of control system: (a) open-loop, (b) closed-loop*

1   *Actuation*
   The set value is used to actuate some element, e.g. a switch, which then adjusts the variable in the process to give the required output.

2   *Process*
   The process is the system in which there is a variable that is being controlled. Thus the process might be a motor with its shaft rotational speed being controlled.

Note that once the set value has been inputted into the open-loop system there is no further control of the output, no signal being fed back from the input to modify the actuation. With a closed-loop system, however, this is not the case as Figure 13.1 illustrates. The basic elements of a closed-loop system are:

1   *Comparison element*
   This compares the required, i.e. set, value of the variable condition being controlled with a signal which is representative of the value that is actually occurring and produces an error signal. It can be regarded as adding the reference signal, which is positive, to the measured value signal, which is negative:

   Error signal = reference value signal – measured value signal

   The feedback uses *negative feedback* because the signal which is fed back subtracts from the input value.

2   *Control element*

This determines the action to be taken when an error signal is received. The control action being used by the element may be just to supply a signal which gives an output which is an on or off signal when there is an error, as in a room thermostat, or perhaps a signal which proportionally opens or closes a valve according to the size of the error.

3   *Correction element*

The correction element acts on the input from the controller to produce a change in the process to correct or change the controlled condition. Thus it might be a switch which switches on a heater and so increases the temperature of the process. The term *actuator* is used for the element of a correction unit that provides the power to carry out the control action.

4   *Process element*

The process is the system in which there is a variable that is being controlled. It could be a room in a house with its temperature being controlled.

5   *Measurement element*

The measurement element produces a signal related to the variable condition of the process that is being controlled. It might be, for example, in a temperature control system a thermocouple which gives an e.m.f. related to the temperature.

13.2.2 Types of closed-loop control systems

Closed-loop control systems can be considered to fall into three main groups:

1   *Systems with continuous control*

The control is continuously exercised with the feedback signal being continuously monitored and compared with the set value. As an illustration of a continuous control system consider the basic elements involved in the control of the speed of a motor by the system shown in Figure 13.2. The position of the slider on a potentiometer is used to set the required speed value by giving a set value voltage. The differential amplifier is used to amplify the difference between this set value and a voltage representing the actual speed and provide an error signal which is used to control the speed of the motor. Basically this might be just a signal which is proportional to the error signal. The speed of the rotating shaft is monitored by a sensor, possibly a tachogenerator, which provides a voltage signal related to the speed of the rotating motor shaft. This signal is then fed back for comparison with the set value signal.

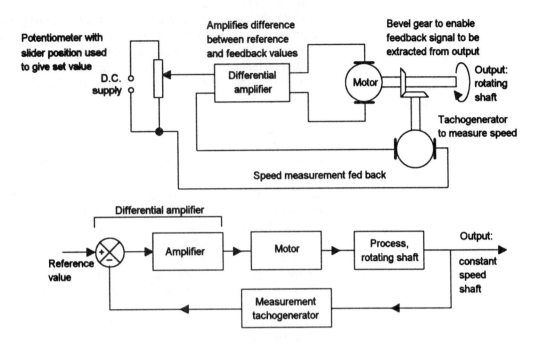

Figure 13.2 *Motor speed control*

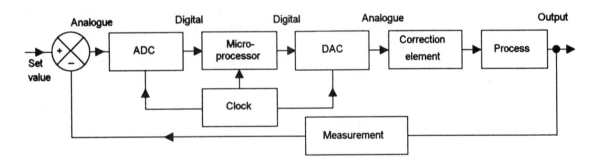

Figure 13.3 *Digital control system*

2 *Systems using digital control which involves sampling*

Digital control, with a system where the error signal is an analogue signal, involves the error being sampled at regular intervals and converted to a digital signal by an analogue-to-digital converter. The control action is then exercised by a microprocessor. Figure 13.3 shows the basic elements involved in such a system. The error between the set value and the actual value, both assumed to be analogue quantities, is converted to a digital signal by an analogue-to-digital converter which samples at regular intervals the analogue error signal. The clock is used to provide the timing signals used to determine when the sampling is to occur. The resulting digital signal is then processed by a microprocessor. This processing involves the microprocessor carrying out calculations based on the error value and possibly stored values of previous inputs and outputs to generate its output signal. Its output is then converted to an analogue signal by a digital-to-analogue converter and used with some correction element to modify the variable in the process system.

3   *Systems using discrete-event control*
This is often termed *sequential control* since the controller is used to sequence a number of discrete events. This involves the control actions being determined by observed sequential conditions or combinations of a set of conditions. For example, a car washer might be controlled so that it comes on when a car is detected as being in the required position under the washing machine and when the correct coins have been inserted in the car wash machine. An automatic machine tool might be controlled to carry out a machining process when sensors detect that the workpiece is in position and then carry out a prescribed sequence of events.

As an illustration of a system using both continuous, or digital, and discrete even control, consider a domestic washing machine (Figure 13.4). Discrete event control is used to switch the various valves and pumps on and off in the required sequence to achieve the selected wash program. However, continuous or digital control is used to control the wash drum rotational speed with the actual drum speed being compared with that required. The following is a typical control program.

1   When the start switch is pressed, open the valves to allow water into the drum.

2   Control the water level and when the full level is reached, close the valves.

3   Switch on the heater.

4   Control the water temperature and when the correct temperature is reached turn on the washer motor.

5   Run the washer motor for a set time.

6   Switch on the pump to empty the drum.

7   When empty, switch off the pump.

8   Open the valves and fill with water.

9   Control the water level and when the full level is reached, close the valves.

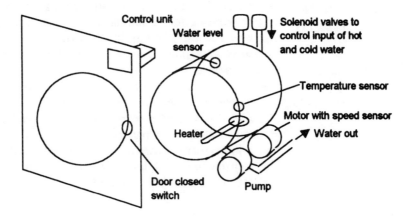

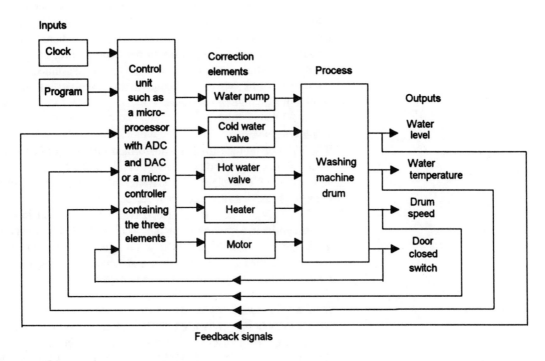

Figure 13.4 *Washing machine system*

10 Start the rinse action of rotating the drum first one way and then the other.

11 Repeat 6, 7, 8, 9 and 10 a number of times.

12 When the drum is empty after rinsing, turn off pump and spin the drum for a set time.

13   End of program.

13.3 Measurement systems

Measurement elements basically consist of a sensor with signal conditioning. *Sensors* are devices which take information about a physical stimulus and turn it into a signal which can be measured, the signal conditioning taking the output from a sensor and converting it into a suitable form for signal processing. The term *transducer* is often used for an element which transforms input signals from some form into an electrically equivalent form. Sometimes the term transducer is used for a combination of a sensor and signal conditioning which takes an input and gives as output an electrical signal. There are a number of ways we can classify sensors, one way being by the principle involved in their operation. The following are some of the commonly used physical principles and examples of sensors.

1   *Resistive sensors*
    The input being measured is transformed into a resistance change. Such sensors include potentiometers where a displacement of the sliding contact results in a change of tapped resistance, thermistors and resistance temperature detectors (RTDs) where a change in temperature results in a change in resistance, strain gauges where a change in strain results in a change in resistance and photo conductive cells where a change in the intensity of illumination results in a change in resistance.

2   *Capacitive sensors*
    The input being measured is transformed into a capacitance change. Such sensors give displacement and pressure sensors where movement of one plate of a parallel plate capacitor results in a change in capacitance, liquid level sensors where the rise in the level of a liquid into the space between two concentric capacitor plates results in a change in capacitance and humidity sensors where a change in humidity results in a change in the amount of water vapour absorbed by the dielectric and hence a change in dielectric constant and consequently a change in capacitance.

3   *Inductive sensors*
    The input being measured is transformed into a change in inductance. A particularly useful form is the linear variable differential transformer (LVDT). This is a a transformer with a primary coil and two identical secondary coils wound on the same former. The displacement of a ferromagnetic core inside the former results in the amount inside one secondary coil increasing while that in the other decreases, as a consequence the inductance of one coil increases and that of the other and the difference is thus a measure of the displacement.

4   *Electromagnetic sensors*
    These are based on Faraday's laws of electromagnetic induction with the input being measured giving rise to an induced e.m.f. The

tachogenerator for the measurement of rotational speed is an example of such a sensor. One form is essentially an a.c. generator with a coil rotating in a magnetic field and giving rise to an alternating current which is a measure of the rate of rotation.

5   *Thermoelectric sensors*
The input is temperature and the output an e.m.f., the sensor being termed a thermocouple.

6   *Elastic sensors*
The input being measured is transformed into a displacement. Examples of such sensors are springs for the measurement of force and diaphragm pressure gauges where the pressure causes a diaphragm to deform, the amount of deformation being used as a measure of the pressure. This diaphragm may be used as one plate of a parallel plate capacitor and so the deformation is transformed into a change in capacitance.

7   *Piezoelectric sensors*
Forces applied to a crystal displace the atoms in the crystal and result in the crystal acquiring a surface charge. Such sensors are used for the measurement of transient pressures, acceleration and vibration.

8   *Semiconductor sensors*
When the temperature of a semiconductor changes, the mobility of the charge carriers change. This affects the rate at which they can move across semiconductor junctions. As a consequence a junction diode can be used for the measurement of temperature, the voltage across it at constant current being a measure of the temperature. Such a sensor with appropriate signal conditioning is supplied as an integrated circuit. Thermotransistors rely on the same basic principle.

9   *Pyroelectric sensors*
Temperature changes give rise to changes in surface charges. Such sensors are widely used for burglar alarm systems to detect the presence of people by their body heat.

10   *Hall effect sensors*
The action of a magnetic field on a flat plate carrying an electric current generates a potential difference which is a measure of the strength of the field.

Sensors can also be classified as passive or active. *Passive sensors* are ones which require an external power supply; *active sensors* are ones which need no such external power supply. A potentiometer is an example of a passive sensor while a thermocouple is an example of an active one.

13.3.1 Terms used to define performance

The following are some of the basic terms used to describe the performance of sensors and measurement systems.

1   *Range*
    The range defines the limits between which the input can vary.

2   *Accuracy*
    Accuracy is the extent to which the value indicated by a measurement system might be wrong. Accuracy is often expressed as a percentage of the full range output or full-scale deflection.

3   *Sensitivity*
    The sensitivity is how much output you get per unit input, e.g. a resistance temperature detector might have a sensitivity of 0.2 Ω/°C. This term is also used to indicate the sensitivity to inputs other than that being measured, i.e. environmental changes. Thus there can be the sensitivity of the transducer to temperature changes in the environment or perhaps fluctuations in the mains voltage supply, e.g. a pressure sensor might have a temperature sensitivity of ±0.1% of the reading per °C change in temperature.

4   *Non-linearity error*
    A linear relationship between the input and output might be assumed over the working range for some sensor when it is not perfectly linear and thus errors occur as a result of the assumption. The error is defined as the maximum difference from the straight line, e.g. a pressure sensor might be quoted as having a non-linearity error of ±0.5% of the full range.

5   *Stability*
    The stability of a transducer is its ability to give the same output when used to measure a constant input over a period of time. The term *drift* is often used to describe the change in output that occurs over time.

6   *Output impedance*
    When a sensor giving an electrical output is interfaced with an electronic circuit it is necessary to know the output impedance since this impedance is being connected in either series or parallel with that circuit and so the inclusion of the sensor can significantly modify the behaviour of the system to which it is connected.

7   *Dynamic characteristics*
    The *static characteristics* are the values given when steady-state conditions occur, i.e. the values given when the transducer has settled down after having received some input, the *dynamic characteristics* being the behaviour between the time that the input value changes and the time that the value given by the transducer

settles down to the steady-state value. Dynamic characteristics are stated in terms of the response of the transducer to inputs in particular forms. Thus the *response time* is the time which elapses after a constant input, a step input, is applied to the transducer up to the point at which the transducer gives an output corresponding to some specified percentage, e.g. 95%, of the value of the input. The *time constant* is the 63.2% response time. A thermocouple in air might have a time constant of perhaps 40 to 100 s. The *rise time* is the time taken for the output to rise to some specified percentage of the steady-state output. Often the rise time refers to the time taken for the output to rise from 10% of the steady-state value to 90 or 95% of the steady-state value.

To illustrate the above, consider the following specifications for temperature sensors:

1 *A bead thermistor*
   Accuracy ±5%
   Maximum power 250 mW
   Dissipation factor 7 mW/°C
   Response time 1.2 s
   Thermal time constant 11 s
   Temperature range –40°C to +125°C

2 *An integrated circuit temperature sensor LM35*
   Accuracy at 25°C ±0.4%
   Non-linearity 0.2°C
   Sensitivity 10 mV/°C
   Quiescent current 65 μA at supply voltage 5 V
   Temperature sensitivity of quiescent current +0.39 mA/°C
   Output impedance 0.1 μΩ with 1 mA load

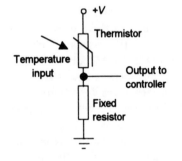

Figure 13.5 *Converting the output of a thermistor to a voltage change*

Signal conditioning is generally necessary with sensors so that their output can be converted into a suitable form for display or use in a control system. For example, the output from a thermistor when there is a temperature change is a change in resistance. This can be converted into a voltage change by incorporating it in a potential divider circuit (Figure 13.5). The resulting voltage change might then be used as the input to a microprocessor used as a controller.

13.4 Electrical switching

Electrical actuators are the elements which are responsible for transforming an electrical output of a controller such as a microprocessor into a controlling action in the process concerned. For example, the electrical output might be transformed into a rotatory or linear motion. Generally there need to be intermediate elements which take the output from the controller, e.g. a microprocessor, and uses it to switch or control a larger current or voltage change which is able to operate the actuation element. The following are examples of such switching elements:

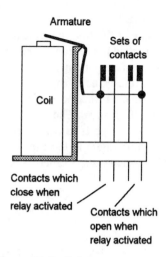

Figure 13.6 *Relay*

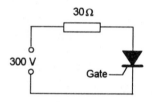

Figure 13.7 *Thyristor circuit*

1 Relays

The electrical relay (Figure 13.6) offers a simple on–off switching action in response to a control signal. When a current flows through the coil of wire a magnetic field is produced. This pulls a movable arm, the armature, that forces the contacts to open or close; usually there are two sets of contacts with one being opened and the other closed by the action. These contacts can then be used to switch on or off a much larger current, e.g. the current through a heater or the current used to operate a motor.

2 Thyristors

A junction diode allows a significant current in one direction through it and barely any current in the reverse direction, i.e. a low resistance in one direction and a high one in the reverse. The *thyristor* or *silicon controlled rectifier* (SCR) can be considered to be a diode which can be switched on to conducting, i.e. switched to a low resistance from a high resistance, at a particular forward direction voltage. The thyristor passes negligible current when reverse biased and when forward biased the current is also negligible until the forward breakdown voltage, e.g. 300 V, is exceeded. Thus if such a thyristor is used in a circuit in series with a resistance of 30 Ω (Figure 13.7) then before breakdown we have a very high resistance in series with the 30 Ω and so virtually all the 300 V is across the thyristor with its high resistance and there is negligible current. When forward breakdown occurs, the resistance of the thyristor drops to a low value and now, of the 300 V, only about 2 V might be dropped across the thyristor. Thus there is now 300 − 2 = 298 V across the 30 Ω resistor and so the current rises from its negligible value to 298/30 = 9.9 A. When once switched on the thyristor remains on until the forward current is reduced to below a level of a few milliamps. The voltage at which forward breakdown occurs is controlled by a gate input, being determined by the current entering the gate, the higher the current the lower the breakdown voltage. Thus by controlling the gate current we can determine when the thyristor will switch from a high to low resistance.

3 Transistors

For the junction transistor in the circuit shown in Figure 13.8(a), when the base current I_B is zero both the base-emitter and the base-collector junctions are reverse biased. When the base current I_B is increased to a high enough value the base-collector junction becomes forward biased. By switching the base current between 0 and such a value, bipolar transistors can be used as switches. When there is no input voltage V_{in} then virtually the entire V_{CC} voltage appears at the output as the resistance between the collector and emitter is high. When the input voltage is made sufficiently high so that the resistance between the collector and emitter drops to a low value, the transistor switches so that very little of the V_{CC} voltage appears at the output (Fig. 7.13(b)). We thus have an electronic switch.

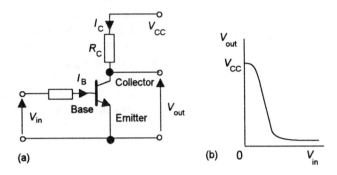

Figure 13.8   *Transistor switch*

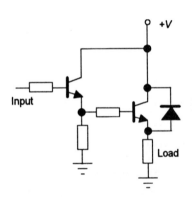

Figure 13.9   *Switching a load*

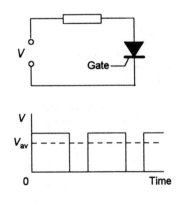

Figure 13.10   *Thyristor d.c. control*

Because the base current needed to drive a bipolar power transistor is fairly large, a second transistor is often needed to take the small current and produce a large enough current to the base of the transistor used for the switching and so enable switching to be obtained with the relatively small currents supplied, for example, by a microprocessor. Such a pair of transistors (Figure 13.9) is termed a *Darlington pair* and they are available as single-chip devices. Since such a circuit is often used with inductive loads and large transient voltages can occur when switching occurs, a protection diode is generally connected in parallel with the switching transistor to prevent damage to it when it is switched off. As an indication of what is available, the integrated circuit ULN2001N contains seven separate Darlington pairs, each pair being provided with a protection diode.

As an illustration of the use of a thyristor, Figure 13.10 shows how it can be used to control the power supplied to a resistive load by chopping a d.c. voltage V. An alternating current signal is applied to the gate so that periodically the voltage V becomes high enough to switch the thyristor off and so the voltage V off. The supply voltage can be chopped and an intermittent voltage produced with an average value which is varied and controlled by the alternating signal to the gate.

Another example of control using a thyristor is that of a.c. for electric heaters, electric motors or lamp dimmers. Figure 13.11 shows a circuit that can be used. The alternating current is applied across the load, e.g. the lamp for a lamp dimming circuit, in series with a thyristor. R_1 is a current-limiting resistor and R_2 is a potentiometer which sets the level at which the thyristor is triggered. The diode in the gate input is to prevent the negative part of the alternating voltage cycle being applied to the gate. By moving the potentiometer slider the gate current can be varied and so the thyristor can be made to trigger at any point between $0°$ and $90°$ in the positive half-cycle of the applied alternating voltage. When the thyristor is triggered near the beginning of the cycle it conducts for the entire positive half-cycle and the maximum power is delivered to the load. When triggering is delayed to later in the cycle it conducts for less time and so the power delivered to the load is reduced. Hence the position of the potentiometer slider controls the power delivered to the load; with the light dimming circuit the slider position controls the power delivered to the lamp and so its brightness.

When the load on a thyristor is inductive, the current flowing through the load does not fall to zero immediately the supply voltage goes negative but decays with time.

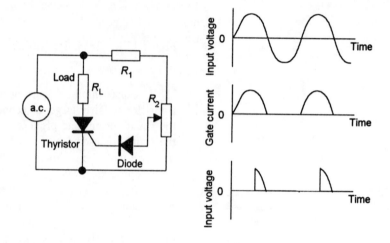

Figure 13.11   *Thyristor control for a.c. power to a load*

13.5 Speed control of motors

Consider the open-loop control of d.c. motor speed by *pulse-width modulation* (PWM). This technique involves the switching on and off of a d.c. voltage to control its average value (Figure 13.12). The greater the fraction of a cycle that the d.c. voltage is switched on the closer its average value is to the input voltage.

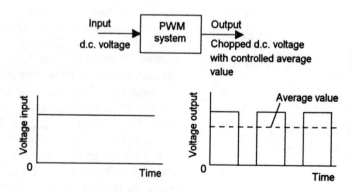

Figure 13.12   *Pulse-width modulation*

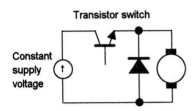

Figure 13.13 *PWM control circuit*

Figure 13.13 shows how pulse width modulation can be achieved by means of a basic transistor circuit. The transistor is switched on and off by means of a signal applied to its base, e.g. the signal from a microprocessor as a sequence of pulses. By varying the time for which the transistor is switched on so the average voltage applied to the motor can be varied and its speed controlled. Because the motor when rotating acts as a generator, the diode is used to provide a path for current which arises when the transistor is off.

Such a basic circuit can only drive the motor in one direction. A circuit (Figure 13.14) involving four transistors, in what is termed an H-circuit, can be used to control both the direction of rotation of the motor and its speed. The motor direction is controlled by which input receives the PWM voltage. In the forward speed motor mode, transistors 1 and 4 are on and current flow is then from left-to-right through the motor. Thus input B is kept low and the PWM signal is applied to input A. For reverse speed, transistors 2 and 3 are on and the current flow is from right-to-left through the motor. Thus input A is kept low and the PWM signal is applied to input B.

Figure 13.15 shows a better version of the H circuit in which logic gates are used to control inputs A and B to achieve the above conditions with now one input supplied with a signal to switch the motor into forward or reverse and the other input the PWM signal. Such a circuit is better suited to microprocessor control for d.c. motors. A high input to the forward/reverse input means that when there is a high PWM signal the AND gate 1 puts transistor 1 on because the two inputs to it are high and so its output is high. The inverter means that AND gate 2 receives a low pulse when the forward/reverse input is high. As a result, transistor 3 is switched off. Because the AND gates 3 and 4 receive the same inputs, transistor 4 is on and transistor 2 is off. The situations are reversed when the signal to the forward/reverse input goes low.

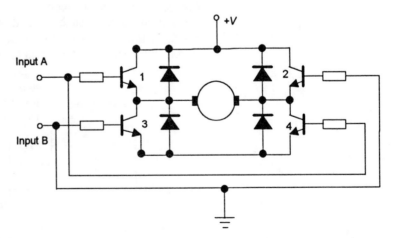

Figure 13.14 *H circuit*

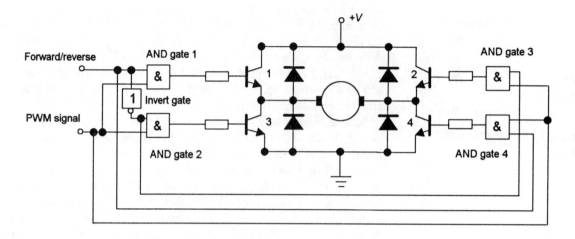

Figure 13.15   *Circuit for microprocessor control of a motor*

The above methods of speed control using PWM have been open-loop systems with the speed being determined by the input to the system and no feedback to modify the input in view of changing load conditions. For a higher grade of speed control than is achieved by the open-loop system, feedback is required. This might be provided by coupling a tachogenerator to the drive shaft (see Figure 13.2). A tachogenerator gives a voltage proportional to the rotational speed. This voltage can be compared with the input voltage used to set the required speed and, after amplification, the error signal used to control the speed of the motor.

Figure 13.16 shows how such a closed-loop system might appear when a microprocessor is used as the controller. The analogue output from the tachogenerator is converted to a digital signal by an analogue-to-digital converter. The microprocessor is programmed to compare the digital feedback signal with the set value and give an output based on the error. This output can then be used to control a PWM circuit and so supply a d.c. signal to the motor to control its speed.

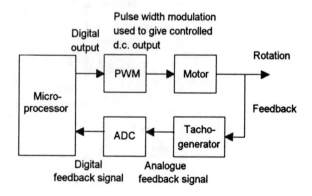

Figure 13.16   *Microprocessor controller with feedback*

13.5.1 Speed control of a.c. motors

The speed of rotation of an a.c. motor depends on the speed of a rotating magnetic field which is determined by the frequency of the a.c. supply to the motor, thus speed control can be achieved by controlling the frequency of this supply. One method of doing this involves first rectifying the a.c. to give d.c. by means of a converter. Then the d.c. voltage is inverted back to a.c. again but at a frequency that can be selected (Figure 13.17).

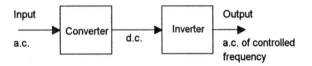

Figure 13.17 *Controlling the speed of an a.c. motor*

A common form of inverter is the H circuit described earlier for pulse width modulation. The supply to it is a d.c. voltage which is chopped to give an on–off voltage output. The output is thus a square wave voltage, the frequency of which can be varied by controlling the frequency with which the input pulses are applied to switch the transistors on or off.   A near-sinusoidal voltage of variable frequency can, however, be produced by varying the duration of each voltage pulse in the way shown in Figure 13.18. This technique is termed *sinusoidal pulse-width modulation*.

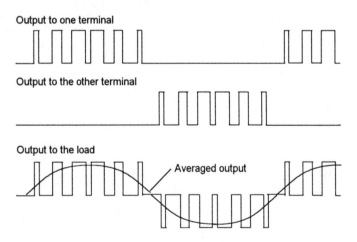

Figure 13.18 *Sinusoidal pulse-width modulation*

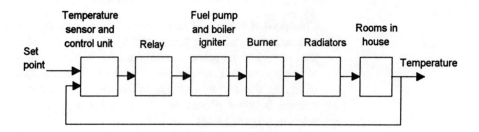

Figure 13.19   *Problem 4*

Problems

1   Explain the difference between open-loop and closed-loop control.

2   Give an example of (a) continuous closed-loop control, (b) open-loop control, (c) sequential control.

3   Explain how (a) relays, (b) thyristors and (c) transistors can be used as switches for actuators in control systems.

4   Figure 13.19 shows a block diagram of a domestic central heating system. Explain its operation.

5   Describe a circuit which uses transistor switching to control the speed and direction of rotation of a d.c. motor.

6   Describe a circuit which could be used as a light dimmer when a light source is connected to the mains power supply.

Answers

Chapter 1
Revision
1   31.8 MPa
2   −8.9 MPa
3   -5×10^{-4}
4   0.56 mm
5   0.30 mm
6   3.6 MPa, 51.7 MPa
7   61.1 MPa, 27.5 MPa, 0.153 mm
8   283.5 kN
9   36.0 MPa, 75.8 MPa
10 68.4 kN
11 6.93 mm
12 0.012 mm, 0.004 mm
13 1.62 MPa
14 64.8 MPa
15 −4 MPa
16 −56 MPa
17 −14.7 MPa, 18.3 MPa
18 −53.0 MPa, 34.0 MPa
19 69.5 MPa, 14.3 MPa

Problems
1   101.9 MPa, 500×10^{-6}
2   163 MPa, 12.2 mm
3   32.4 MPa, 0.24 mm
4   3.0 mm
5   20 mm
6   0.20 mm
7   64.6 MPa, 38.8 MPa
8   93.7 MPa, 6.6 MPa
9   234 kN
10 76 mm
11 1.80 mm, -5.41×10^{-3} mm
12 67.2 MPa
13 180 MPa
14 30.8°C, 50 MPa
15 22.8 MPa, −41.5 MPa
16 93.3 MPa, −68.9 MPa

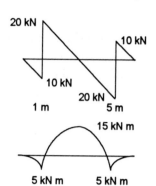

Figure A.1   *Revision problem 4, Chapter 2*

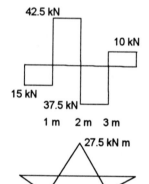

Figure A.2   *Revision problem 5, Chapter 2*

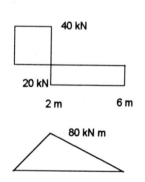

Figure A.3   *Problem 4, Chapter 2*

Chapter 2

Revision

1   (a) –250 N, + 250 N m, (b) +250 N, + 375 N m
2   (a) +8 kN, –12 kN m, (b) +8 kN, –8 kN m
3   +20 kN, – 20 kN m
4   See Figure A.1, 15 kN m at midpoint
5   See Figure A.2, 27.5 kN m at midpoint
6   0.96 mm
7   97 mm from base
8   317.7×10^6 mm^4
9   339 kN m
10   720 N m
11   4.8 kN m
12   2.95×10^{-4} m^3

Problems

1   (a) –10 kN, + 5kN m, (b) –10 kN, +10 kN m
2   (a) +15 kN, +8.75 kN m, (b) +10 kN, +17.5 kN m
3   As in Figure 2.14, maximum bending moment 60 kN m at midpoint
4   See Figure A.3, 80 kN m at 2 m
5   20 kN, –20 kN m
6   See Figure A.4, maximum shear force 60 kN at fixed end, maximum bending moment 130 kN m at fixed end
7   See Figure A.5, maximum shear force 78 kN, 8 m from supported end, maximum bending moment –112 kN m, 8 m from supported end
8   420 MPa
9   89.5 mm from base
10   4.17×10^6 mm^4
11   76.3 mm from base, 2.03×10^6 mm^4

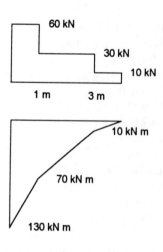

Figure A.4   *Problem 6, Chapter 2*

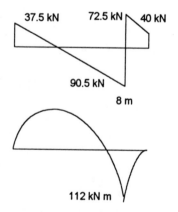

Figure A.5   *Problem 7, Chapter 2*

12   15 kN m
13   ±130 MPa
14   3.57×10^{-3} m^3
15   (a) 55 mm from bottom, 5.89×10^6 mm^4, 1.07×10^5 mm^3, (b)
      30 mm from bottom, 5.54×10^5 mm^4, 1.85×10^4 mm^3, (c) 15 mm
      from bottom, 2.08×10^5 mm^4, 5.95×10^3 mm^3, (d) 22.5 mm from
      bottom, 8.61×10^5 mm^4, 2.30×10^4 mm^3
16   26 MPa

Chapter 3

Revision
1   14.6 mm
2   12.6 mm
3   141 kN
4   0.01 mm
5   2.5×10^{-4}
6   1.7 kN m
7   1.8 kN m
8   0.127 m
9   150 mm
10   33.3 MPa
11   15.8 MPa
12   501 kW

Problems
1   2.5 kN
2   0.006 mm
3   22 kN
4   19.9 MPa
5   49.6 mm
6   1.26 kN m
7   (a) 65.2 MPa, (b) 149.4 GPa, (c) 4.36×10^{-4}
8   (a) Tubular shaft stress is 1.15 that in solid, (b) tubular shaft angle is
      1.15 that in solid, (c) tubular shaft mass is 0.64 that of solid
9   53.1 MPa
10   Tubular shaft torque is 1.44 that of solid shaft
11   33.3 MPa, 46.9 MPa
12   165 MPa
13   10 kN m
14   17.7 Hz
15   4.4 mm
16   (a) 45.0 MPa, (b) 4.3°
17   257.8 N m, 81.0 kW
18   290 kW

Chapter 4

Revision
1   128 m
2   40 m
3   7.5 s, 112.5 m

4   (a) −2 m/s², (b) 2 m
5   2.0 s
6   11.5 m
7   27.6 m
8   23.5 km/h 70° east of north
9   1.37 m/s east, 3.76 m/s north
10   1.27 m, 8.83 m
11   31.0°, 59.0°
12   13.2 m, 3.28 s, 62.7 m
13   8.64 rad/s², 25
14   0.21, 4.7
15   21.2
16   72.7 rad/s
17   20 rad/s²
18   21.2 rad/s, 2.67 rad/s²
19   400 mm
20   5.75×10^{-3} m/s
21   23.5 N, 1.96 m/s²
22   5.12 m/s²
23   0.58
24   2.19 m/s²
25   16.2 N
26   (a) 520 N, b) 980 N
27   0.16 kg m²
28   4.29 kg m²
29   1.196 kg m²
30   11.8 kg m²
31   0.375 N m
32   102 N
33   3.27 m/s²
34   89 rad/s
35   157 J
36   20 kW
37   14.0 rad/s
38   3.1 rad/s
39   137 N

Problems
1   15 m/s
2   16 m
3   5 s
4   (a) 16 m, (b) 7 m
5   16 m, 4 s
6   40 m/s
7   4 ms, 500 m/s²
8   61.2 s
9   50 s
10   2 m/s²
11   (a) 26 m/s, (b) 8 m/s²
12   22.6 m

13 (a) 9.1 m, (b) 8.4 m
14 25 m/s 16° east of north
15 5 m/s at 36.9° to the horizon
16 1.4 m/s north, 1.4 m/s east
17 40.8 m
18 16.3 km
19 20.7 m/s
20 17.1 m
21 28.1 km
22 0.84 rad/s², 19.2 revs
23 0.31 rad/s², 7.5 revs
24 1.26 rad/s²
25 2.1 m/s
26 (a) 0.57 rad/s², (b) 0.21 m/s²
27 8.3 rad/s²
28 300 mm
29 7.5 rad/s²
30 100 rev/min
31 2.39 m/s², 24.4 N
32 3.68 m/s², 18.4 N
33 27.2 kg
34 As given in the problem
35 1.67 m/s², 1300 N
36 $\sqrt{(2d\{M + m\}/Mg)}$
37 0.052
38 0.41
39 0.24, 22.9 N
40 As given in the problem
41 0.02 kg m²
42 1.35 kg m²
43 0.5 m
44 22.8 kg m²
45 45.9 kg m²
46 14.8 rad/s
47 24.5 kg m²
48 20 rad/s²
49 135.7 N m
50 As given in the problem
51 2.4 s
52 4.9 rad/s

Chapter 5   *Revision*
1   0, 10.1 m/s, 2.53×10^3 m/s², 0
2   5 m/s
3   26.6 N
4   2.22 J
5   2.23 Hz
6   2.25 Hz

7   $\dfrac{1}{2\pi}\sqrt{\dfrac{9}{\left(\dfrac{4}{k_1}+\dfrac{1}{k_2}\right)M}}$

8   11.3 Hz

9   1.23 Hz

10 Period increases to 1.001 s

11 0.59 Hz

Problems

1   3.77 m/s

2   2.25 Hz

3   3.0 m/s, 56.8 m/s^2

4   4.44 s

5   3.1 m/s, 2.7 m/s

6   3.87 J

7   790 N/m

8   1.125 Hz

9   2.25 Hz

10 27.6 Hz

11 1.29 Hz

12 Period increases to 1.002 s

13 1.59 Hz

14 3.53 Hz

15 As given in the problem

16 As given in the problem

17 0.040 m

18 $\dfrac{1}{2\pi}\sqrt{\dfrac{2k}{m}}$

19 (a) 2.5 s, (b) 12.1 s

20 Reduced by factor of $1/\sqrt{2}$

21 24 mm, 19 mm, 16 mm

Chapter 6   *Revision*

1   3.6 kW

2   645 W

3   0.25 K/W

4   25.7 W

5   56 W

6   (a) Parallel resistors, (b) as in Figure 6.6

7   33 W

8   166 W

9   4.620 K/W

10 0.14 K/W

11 Rate of heat loss from larger diameter pipe 16 times greater than that from smaller diameter

12 17.8 W

13 4.95 kW

14 185 W

15 8.7 W

16   50.1 W

Problems
1   20 kW
2   76 W, the lagging prevents transverse heat conduction
3   157 W
4   60.6 W
5   3 kW
6   18.7 mm
7   (a) Resistors in series, (b) as in Figure 6.6
8   0.22
9   23.5 W
10   0.16 K/W
11   Increase convective heat loss by increasing h and area
12   112.6 W
13   23.4 W
14   1088 K
15   281 W
16   14.0 W

Chapter 7   *Revision*
1   120 Pa
2   2/3
3   56.5 N, 28.2 W
4   34.1 W
5   46.5 W
6   0.02 m/s
7   1529 m of oil
8   Turbulent, 17.3 kPa
9   136.6 m of water
10   159 kW

Problems
1   1.0×10^{-6} m^2/s
2   0.27 m/s
3   5000 Pa
4   30.2 N, 12.1 W
5   150.7 W
6   17.9 W
7   355 W
8    $T = \pi \eta \omega d^4/16s$
9   431 W
10   13.6 m/s
11   18.8 m of oil
12   Laminar, 17.0 m of oil
13   Laminar 8.0 m of oil
14   55.5 kW
15   0.131 W
16   2.05 kW

Chapter 8   *Revision*

1   5.4 V at 21.8° leading 5 V phasor
2   3.6 A at 56.4° leading 2 A phasor
3   10.8 V at 21.8° lagging from sum phasor
4   2500 Ω, 0.0004 S
5   19.9 Ω
6   1 kΩ, 16 sin 2000*t* mA
7   32 Ω, 3.2 sin 400*t* V
8   500 Ω, 53.1°
9   134 mA, −26.6°
10   3805 mA, −29.9°
11   2.88 mA, 87.3°
12   103.1 Ω, −14.0°
13   0.400 A, 72.6° lagging, 30 Ω
14   14.2 mA, 32.1° leading, 845 Ω
15   30.8 µF
16   576 Ω
17   1.14 W
18   0.124 W
19   82.99 W, 154.5 V A
20   0.6 lagging, 4.61 kV A, 3.69 V Ar, 2.77 kW
21   244 kV Ar
22   35.4 µF

Problems

1   (a) 1592 Ω, (b) 2 Ω, (c) 4167 Ω
2   (a) 2513 Ω, (b) 4000 Ω, (c) 10 Ω
3   0.4 sin 1000*t* mA
4   26.5 kHz
5   79.6 Hz
6   21.7 Ω, 11.1 A, 46.3° current lagging voltage
7   17 Ω, 28.1° voltage leading current
8   (a) 75 Ω, (b) 1.33 A, (c) 59.9 V, 79.8 V
9   (a) 24.1 Ω, (b) 4.15 A at 51.4° lagging voltage
10   148 µF
11   (a) 75.2 Ω, (b) 3.19 A, 57.9° leading voltage
12   (a) 48.0 Ω, (b) 0.5 A, 33.5° leading voltage
13   (a) 70.7 V, 45° current lagging voltage, (b) 50 V, 100 V, 50 V
14   (a) 0.86 A, 49.9° lagging voltage, (b) 174 V, (c) 86 V
15   28.4 A, 32.5° lagging the voltage
16   5 A, 53.1° lagging the voltage
17   200 Ω, 5.3 µF
18   (a) 1.73 A, 1.0 A, (b) 57.7 Ω, 100 Ω
19   39.8 mV
20   0.099 leading, 22.8 V A, 22.69 V A r, 2.26 W
21   (a) 38.7 mA, 83.1° lagging, (b) 0.12, (c) 1.11 W, (d) 9.28 V A, (e) 9.21 V A r
22   (a) 0.53, (b) 8.15 W, (c) 15.3 V A, (d) 10.8 V Ar
23   133.3 kV A, 106.7 kV Ar
24   112.5 kV Ar

25   24.9 µF

Chapter 9

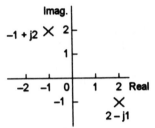

Figure A.6   *Revision problem 1, Chapter 9*

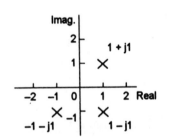

Figure A.7   *Problem 1, Chapter 9*

Revision

1   See Figure A.6
2   (a) $5\angle127°$, (b) $3\angle0°$, (c) $5\angle53°$, (d) $5\angle233°$, (e) $4\angle270°$
3   (a) $2.5 + j4.3$, (b) $1.7 - j1$, (c) $-3.5 - j2$, (d) $3.9 - j4.6$
4   (a) $3 + j8$, (b) $5 + j1$, (c) $7 + j6$, (d) $1 + j2$, (e) $-3 + j5$
5   (a) $22 - j14$, (b) $11 - j2$, (c) $10 + j5$
6   (a) $10\angle40°$, (b) $2\angle40°$, (c) $3\angle120°$
7   (a) $3 + j5$, (b) $-2 - j2$, (c) $-4 + j6$
8   (a) 5, (b) 20, (c) 13
9   (a) $-0.24 + j0.68$, (b) $4 + j1$, (c) $0.5 + j0.5$, (d) $0.14 + j0.66$
10 (a) $2\angle50°$, (b) $2.5\angle(-50°)$, (c) $0.1\angle10°$, (d) $0.5\angle140°$
11 (a) $10\angle0°$ V, 10 V, (b) $2\angle60°$ A, $1 + j1.73$ A, (c) $5\angle90°$ V, j5 V
12 (a) $3 - j2$ V, $3.6\angle(-33.7°)$ V, (b) $2 - j5$ V, $5.4\angle(-68.2°)$ V,
     (c) $5.5 + j2.6$ V, $6.1\angle25.3°$ V, (d) $12.1 + j7.1$ V, $14.0\angle30.4°$ V
13 $1.45 \sin(314t + 45.5°)$ A
14 (a) $36 + j2$ W, (b) $8\angle90°$ W, (c) $8\angle30°$ W
15 (a) $-0.8 + j1.4$ Ω, $3\angle20°$ Ω, (c) $5\angle(-30°)$ Ω
16 (a) $16\angle90°$ Ω, j16 Ω, (b) $0.16\angle130°$ V
17 (a) $20 + j100$ Ω, (b) $100 - j40$ Ω, (c) $10 + j15$ Ω
18 $0.134\angle(-26.6°)$ A
19 $4.73 \sin(240t + 74.9°)$ mA, $15.5 \sin(240t + 74.9°)$ V,
     $8.96 \sin(240t - 15.2°)$ V
20 (a) $4 + j8$ Ω, (b) $100 - j100$ Ω
21 (a) $0.01 + j0.001$ S, (b) $0.002 - j0.005$ A
22 $0.5 - j0.3$ S, $1.47 + j0.88$ Ω
23 $0.02 - j0.0125$ S, $1.42\angle(-32°)$ A or $1.2 - j0.75$ A
24 $6.93 + j5.38 = 8.77\angle37.8°$ Ω
25 $2.0\angle36.9°$ A

Problems

1   See Figure A.7
2   (a) $2.2\angle166°$, (b) $5\angle233°$, (c) $3\angle0°$, (d) $6\angle270°$, (e) $1.4\angle45°$,
     (f) $3.6\angle326°$
3   (a) $-2.5 + j4.3$, (b) $7.07 + j7.07$, (c) -6, (d) $0.68 + j2.72$,
     (e) $1.73 + j1$
4   (a) $1 + j6$, (b) $5 - j2$, (c) $-14 + j8$, (d) $0.23 - j0.15$, (e) $0.1 - j0.8$
5   (a) $5 - j2$, (b) $-2 - j1$, (c) $-1 + j7$, (d) 1, (e) $12 + j8$, (f) $-10 + j6$,
     (g) $11 - j2$, (h) $10 - j2$, (i) $0.9 + j1.2$, (j) $0.23 - j0.15$, (l) j1,
     (m) $-0.3 + j1.1$
6   (a) $20\angle60°$, (b) $50\angle80°$, (c) $0.1\angle(-20°)$, (d) $0.5\angle(-40°)$,
     (e) $5\angle(-20°)$, (f) $0.4\angle(-20°)$
7   (a) $10\angle(-30°)$, $8.66 - j5$, (b) $10\angle150°$, $-8.66 + j5$, (c) $22\angle45°$,
     $15.6 + j15.6$
8   (a) $5.5 + j2.6$, $6.1\angle25.3°$, (b) $-2 + j7$, $7.3\angle105.9°$, (c) $3.7 + j4.5$,
     $5.8\angle50.6°$

9 (a) $25\angle90°$, (b) $20\angle75°$, (c) $44.5\angle83.3°$, (d) $4\angle(-30°)$, (e) $1.25\angle15°$, (f) $0.164\angle9.2°$

10 (a) $20 + j17.32 = 26.46\angle40.9°$ V, (b) $26.46\sin(\omega t + 40.9°)$ V

11 $25\angle(-30°)$ Ω

12 $2\angle(-36.8°)$ A

13 (a) $318\angle(-90°)$ Ω, $j200$ Ω, (b) $1.3\sin(314t - 50°)$ mA

14 (a) $200\angle90°$ W, $j200$ W, (b) $60\sin(2000t + 110°)$ mA

15 (a) $12 - j5$ Ω, (b) $136.6 + j136.6$ Ω, (c) $32.1 + j7.4$ Ω, (d) $1.88 - j6.34$ Ω, (e) $0.384 - j1.922$ Ω, (f) $j13.3$ Ω

16 (a) $5 + j2$ Ω, (b) $50 - j10$ Ω, (c) $2 + j1$ Ω, (d) $1.92 - j0.38$ Ω, (e) $-j$ 125 Ω

17 (a) $0.05 - j0.05$ S, (b) $0.1 + j0.0025$ S, (c) $0.005 + j0.00042$ S

18 $0.0884\sin(314t - 45°)$ A

19 $0.24\sin\omega t$ A, $0.24\sin(\omega t + 90°)$ A

20 $2.83\sin(\omega t + 45°)$ A

21 (a) $3.67\angle(-40.9°)$ A, (b) $1.79\angle63.4°$ A

22 $7.81\angle(-38.7°)$ V, $9.38\angle(-128°)$ V, $15.63\angle51.3°$ V

23 $85.8\angle(-5.7°)$ V

24 $8.72 + j4.75 = 8.84\angle28.8$ Ω, $11.3\angle(-28.6°)$ A

25 $1137.9 - j944.7 = 1479\angle(-39.7°)$ Ω, $33.8\angle39.7°$ mA

Chapter 10

Revision

1 5033 Hz, $0.25\angle0°$ A

2 91.9 Hz, $2.5\angle0°$ A

3 205 Hz, $129\angle(-90°)$ V

4 21.1 Ω

5 1989 Hz, 100 Ω, $60\angle0°$ mA

6 4490 Hz, 2000 Ω, $5\angle0°$ mA

7 159 Hz, 100

Problems

1 1125 Hz, $1.2\angle0°$ A

2 356 Hz

3 $15\angle0°$ V, $300\angle90°$ V, $300\angle(-90°)$ V

4 (a) 8.4 μF, (b) 9.4

5 (a) 118.6 Hz, (b) 11.2

6 7118 Hz

7 2251 Hz, 100 kΩ, 1 mA, 141

8 57.8 Hz, 200 Ω, 0.5 A, 4.6

Chapter 11

Revision

1 Second and fourth (and a d.c. term)

2 (a) See Figure A.8, (b) sum graph of Figure A.8 lifted up by 5

3 (a) Only even harmonics, (b) only odd harmonics

4 $0.32 + 0.5\cos 100t + 0.21\cos 200t$ mA

5 $0.5\cos(100t - 90°) + 0.21\cos(200t - 90°)$ A

6 $3.2\sin(100t + 90°) + 3.2\sin(200t + 90°)$ mA

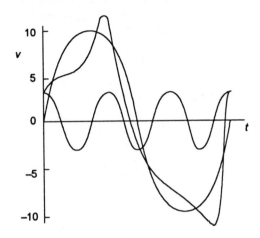

Figure A.8 *Revision problem 2, Chapter 11*

7    $5.32 \cos(314t + 57.9°) - 1.56 \sin(2 \times 314t + 38.5°)$ A
8    $0.63 \sin(500t + 71.6°) + 0.21 \sin(1500t - 59.0°)$ mA
9    $-1.40 \cos(400t + 56.3°) - 0.12 \cos(1200t - 65.4°)$
     $- 0.022 \cos(2000t - 77.5°)$ A
10   (a) 530.5 Hz, (b) $0.0038 \sin(3333t + 89.98°) + 0.4 \sin 1000t$ A
11   15%
12   (a) 72 mV, (b) 2.1%

Problems
1   See Figure A.9
2   See Figure A.10
3   (a) Only odd harmonics, (b) only even harmonics
4    $0.2 \sin 500t + 0.1 \sin 1000t$ A
5    $0.2 \sin(500t + 53.1°) + 0.14 \sin(1000t + 33.7°)$ A
6    $5 \sin(\omega t + 36.9°) + 2.89 \sin(3\omega t - 37.4°)$ A
7    $2.8 \sin(500t - 45°) + 1.1 \sin(1500t - 18.4°)$ A

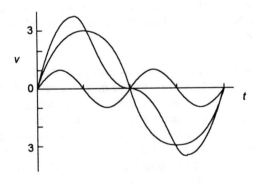

Figure A.9   *Problem 1, Chapter 11*

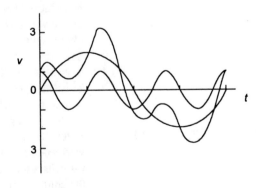

Figure A.10   *Problem 2, Chapter 11*

8   $5 + 14.1 \sin(10\,000t - 45°) + 2.53 \sin(30\,000t - 41.6°)$ A
9   (a) 118.6 Hz, (b) $0.0105 \sin(745t + 89.88°) + 0.4 \sin 3 \times 745t$ A
10 1007 Hz
11 43%
12 10%

Chapter 12   *Revision*
1   9
2   33.3, 221 762
3   7.94
4   30 dB
5   17 dB
6   16.4
7   10^{-4} W

Problems
1   1.22 mV
2   8
3   0.625 V
4   18 mV
5   19.95
6   15 dB
7   19 dB
8   500
9   0.008
10 41.2 dB

Chapter 13   *Problems*
1   Open-loop has no feedback and so output determined solely by set point setting. Closed-loop control has feedback and so can react to changes on the output variable, e.g. loading.
2   For example you might have: (a) thermostatically controlled heating system, (b) an electric fire which is purely controlled by the set point determined by how many bars are switched on, (c) a washing machine where one event such as water at temperature is followed by the washing motor being switched on, this in turn being followed after a set time by another operation.
3   See Section 13.4
4   It is a closed-loop system. The control unit compares the set and actual temperature values and gives a signal to operate a relay. This is able to switch on a larger current and hence the fuel pump and igniter. This gives an input to the boiler of heat and an output of hot water. This is then the input to radiators which give an output of warmed air to the process which are the rooms in the house. The result is a temperature output. The temperature signal is fed back to the control unit.
5   See Figure 13.13
6   See Figure 13.10

Index